Climate Change and Development

This book addresses the two greatest challenges of our time – averting catastrophic climate change and eradicating poverty – and the close interconnections between them. The evidence that humans are causing climate change is now overwhelming, the brunt of its impacts is already being felt by poor people, and the case for urgent action is compelling. This text provides a comprehensive and multi-disciplinary foundation for understanding the complex and tangled relationship between development and climate change. It argues that transformational approaches are required in order to reconcile poverty reduction and climate protection and secure sustained prosperity in the twenty-first century.

Section One of the book provides the building blocks for understanding climate science and the nexus between climate and development. Section Two outlines responses to climate change from the perspective of developing countries, with chapters on international agreements, climate change mitigation and adaptation, and climate finance. Each chapter offers analytical tools for evaluating responses, enabling readers to ask smart questions about the relationship between climate change and development as policy and action evolve in the coming years. The last three chapters of the book, contained in Section Three, are forward looking and focus on why and how development must be re-framed to deliver more equitable and sustainable outcomes. It sets out different critiques of 'development as usual' and explores competing visions of development in a warming and resource-constrained world.

Climate Change and Development uses real-world examples to bring together perspectives from across different disciplines. This invaluable and clearly written text contains start-of-chapter learning outcomes, end-of-chapter summaries, discussion questions, suggestions for further reading and relevant websites. The text is suitable for both undergraduate and postgraduate students, as well as those working in international development contexts who wish to get to grips with this pressing global challenge. As the state of knowledge and of action in both fields is rapidly evolving, readers are referred to the accompanying website www.climateandevelopment.com for updates on latest developments. This also serves as a general resource on the climate-development nexus.

Thomas Tanner is a Research Fellow at the Institute of Development Studies (IDS), UK. He is a development geographer specializing in poverty, sustainability and climate change adaptation in the context of international development. He has extensive experience as a researcher, policy maker, project manager and a UN negotiator on climate change and other environment and development issues.

Leo Horn-Phathanothai is Director for International Cooperation at the World Resources Institute. He is an environmental economist with a rich background in international development diplomacy. He has worked in the British and international civil services, providing strategic and policy advice to the British and Chinese governments, and international organizations, including the World Bank and the United Nations Development Programme.

The challenges of underdevelopment and climate change have the same roots and potentially the same solutions. This accessible, clear and insightful analysis spells out in the greatest detail the moral, political and technical elements of both. Highly recommended.
Neil Adger, University of Exeter

This book presents a critical analysis of the relationship between climate change and development. The authors convincingly argue that current development practices, policies and pathways need to be transformed to meet the challenges of climate change. They also provide exciting examples of emerging approaches and new paradigms for sustainable development that will be of interest to both students and practitioners.
Karen O'Brien, University of Oslo

As climate change is becoming recognized as the defining issue of our era, this book is essential reading for both students and professionals in NGOs, the UN and other agencies working on the interface between environment and development.
Dr Saleemul Huq, Director, International Centre for Climate Change and Development, Independent University, Bangladesh

Finding the right way forward to simultaneously address both the world's pressing development needs and the equally pressing need to save the planet's climate - that's the most important task of our time. Climate Change and Development is the best text available on this challenging, vital subject. Bringing equity and environment together isn't easy, but this book is a major contribution in this regard. Highly recommended.
James Gustave Speth, author of America the Possible: Manifesto for a New Economy, *and former Administrator, United Nations Development Programme*

Routledge Perspectives on Development

Series Editor: Professor Tony Binns, *University of Otago*

Since it was established in 2000, the same year as the Millennium Development Goals were set by the United Nations, the *Routledge Perspectives on Development* series has become the pre-eminent international textbook series on key development issues. Written by leading authors in their fields, the books have been popular with academics and students working in disciplines such as anthropology, economics, geography, international relations, politics and sociology. The series has also proved to be of particular interest to those working in interdisciplinary fields, such as area studies (African, Asian and Latin American studies), development studies, environmental studies, peace and conflict studies, rural and urban studies, travel and tourism.

If you would like to submit a book proposal for the series, please contact the Series Editor, Tony Binns, on: jab@geography.otago.ac.nz

Published:

Third World Cities, 2nd edn
David W. Drakakis-Smith

Rural–Urban Interactions in the Developing World
Kenneth Lynch

Environmental Management & Development
Chris Barrow

Tourism and Development
Richard Sharpley and David J. Telfer

Southeast Asian Development
Andrew McGregor

Population and Development
W. T. S. Gould

Postcolonialism and Development
Cheryl McEwan

Conflict and Development
Andrew Williams and Roger MacGinty

Disaster and Development
Andrew Collins

Non-Governmental Organisations and Development
David Lewis and Nazneen Kanji

Cities and Development
Jo Beall

Gender and Development, 2nd edn
Janet Momsen

Economics and Development Studies
Michael Tribe, Frederick Nixson and Andrew Sumner

Water Resources and Development
Clive Agnew and Philip Woodhouse

Theories and Practices of Development, 2nd edn
Katie Willis

Food and Development
E. M. Young

An Introduction to Sustainable Development, 4th edn
Jennifer Elliott

Latin American Development
Julie Cupples

Climate Change and Development

Thomas Tanner and
Leo Horn-Phathanothai

Routledge
Taylor & Francis Group

LONDON AND NEW YORK

First published 2014
by Routledge
2 Park Square, Milton Park, Abingdon, Oxon OX14 4RN

and by Routledge
711 Third Avenue, New York, NY 10017

Routledge is an imprint of the Taylor & Francis Group, an informa business

© 2014 Thomas Tanner and Leo Horn-Phathanothai

British Library Cataloguing in Publication Data
A catalogue record for this book is available from the British Library

Library of Congress Cataloging in Publication Data
Tanner, Thomas, 1975– author.
Climate change and development / Thomas Tanner and Leo Horn-Phathanothai.
pages cm. — (Routledge perspectives on development)
Includes bibliographical references and index.
1. Economic development—Environmental aspects.   2. Sustainable development.
3. Climatic changes.   I. Horn-Phathanothai, Leo, author.   II. Title.
HD75.6.T375 2014
338.9'27—dc23
2013021206

ISBN: 978–0–415–66426–4 (hbk)
ISBN: 978–0–415–66427–1(pbk)
ISBN: 978–0–203–81886–2 (ebk)

Typeset in Times New Roman and Franklin Gothic
by RefineCatch Limited, Bungay, Suffolk

Contents

Plates

 **Figures**

 **Tables**

Boxes

# Foreword

The reality of a world with more extreme weather events, rising seas, and longer droughts is becoming clearer by the day. With every new season new records are being set: as I write we are currently in the 340th consecutive month with a global temperature above the 20th-century average.

Sadly, the global economy remains wedded to fossil fuel energy that is driving climate change. Currently, governments spend roughly six times more on subsidies for fossil fuels than renewable energy – and that's despite many fossil fuel companies being among the largest and most profitable in the world. With one hand governments are encouraging households and businesses to use less fossil fuel, and with the other are effectively paying them to use more! Recent analysis from WRI finds that there are nearly 1,200 new coal plants under planning worldwide. If these plants come online, our chances of staying within 2 degrees of warming – the level recommended to prevent the worst consequences of climate change – would be almost nil.

As we continue on the path of increasing emissions we are heading in to dangerous, unchartered waters. The World Bank, an organization not prone to hyperbole, warned that the prospects of a 4 degree rise in global temperatures is real, and would have devastating consequences.

The impacts of climate change are profoundly unfair and tackling it raises profound questions of equity and justice. The loss and suffering wrought by Hurricane Sandy pales in comparison to the impacts of weather related catastrophes in the developing world. The causes and effects of climate change are not evenly distributed – the poorest and most vulnerable, who have contributed the least to the problem, stand to suffer the most from climate change impacts. In a warming world, vulnerable communities would face even more food disruption and water scarcity, along with more diseases and pests.

This book addresses the two biggest challenges of this century – stabilizing the climate and eradicating poverty – and the complex linkages between them. The authors note that climate protection and development objectives are deeply intertwined, yet are still being advanced largely in parallel by professional communities and agencies that speak different languages. This fragmentation of knowledge and action has bred deeply misguided notions that climate action and development are separate concerns at best, if not conflicting interests. By bringing together insights from economics, geography, climate science and development studies, the authors help build the kind of knowledge base that is needed to navigate the complex interactions between climate change and development.

While emissions trends are pointing in the wrong direction there is cause yet for optimism as we look to the future. The world has made remarkable strides in reducing poverty, spreading education, improving maternal health and checking the spread of HIV over the past few decades, and we have learnt much in the process about how to bring transformative solutions to scale. So too we are beginning to see a modest gain in momentum on climate action: more than 90 countries now have goals for helping to address climate change, investments in renewable energy, at around $300 billion last year are up 8-fold in the past decade, and around the world private companies, citizens' groups and local authorities are demanding clearer climate change policies from their governments. Progress is real, but to date the scale and pace of such progress is simply not commensurate with the scale and urgency of the challenge. The challenge is to accelerate this progress in the very short time that we have at our disposal so that total greenhouse gas emissions stop rising before 2020, and begin to decline at an accelerating pace through the subsequent decade. By 2050 total emissions in rich countries will need to be 80–90 per cent lower than in 2000.

If this tipping point is to be crossed, citizens, politicians and businesses need to realize that it is their own advantage. The good news is that there is growing evidence that smart environmental and growth policies combined can actually promote efficiency gains and technological advances, increase investment and generate competitive advantage. Evidence suggests that it is possible to decouple environmental damage from growth, decarbonize production, use resources more efficiently, farm differently and design smarter urban infrastructure at a reasonable cost (1–3 per cent of GDP per year). In sum, 'green' and 'growth' are not mutually incompatible, as they have so often been regarded.

The development community has only in recent years started waking up to the development risks and opportunities linked to climate change. Although many of the root causes have been known for years, and many of the solutions exist, our progress to date has been too little and too slow. This is in part a failure of my generation. I am hopeful as I look at the next generation that will be picking up the baton. This textbook provides an important reference and guide for anyone serious about addressing the climate challenge and realizing the promise – that lies within our reach – of a world rid of poverty and hunger, in balance with nature.

Dr Andrew Steer
President and CEO of The World Resources Institute (WRI)

Acknowledgements

This book would never have seen the light of day were it not for the unflinching support and encouragement of our families and friends, and the knowledge and inspiration of countless colleagues. First and foremost, we are deeply grateful to our wonderful partners, Gemma and Libbet, who saw us through many late nights and book-filled weekends as well as providing invaluable feedback on style and content. Thanks also to Brian and Ruth Tanner, Sirin Phathanothai, David Horn and Anton Smitsendonk for inspiring us to take it on.

We benefited greatly from excellent review comments. Thanks in particular go to Joe Horn, Heather McGray, Andrew Steer, Anton Smitsendonk, Jonathan Moore, Karnika Palwa, Emily Schabacker, Sarah Parsons, Kevin Rowe, Gareth Elston and Nadine Beard. We wish to thank also the anonymous reviewers that shaped the initial design and the first draft. Thank you to Faye Leerink and Andrew Mould at Routledge for your professionalism in guiding us through the process, and series editor Tony Binns for helpful and encouraging early feedback.

Special thanks go to Nicolas Douillet (http://nicdouillet.tumblr.com) for his cartoons, which provide a fantastic visual opener to each of the chapters. Thanks also to authors, publishers and photographers that granted us permissions to reproduce their work in this book.

The list is long of colleagues and mentors to whom we owe much of what we have learned and distilled in this book. In particular we would like to acknowledge the following generous souls: Aditya Bahadur, Manish Bapna, Steve Bass, Terry Cannon, Declan Conway, Gordon Conway, Angie Dazé, Tim Foy, Anne Hammill, Saleemul Huq, Ma Zhong, Tom Mitchell, Lars Otto Naess, David Norse, David Ockwell, Janet Ranganathan, Fran Seballos, Paul Steele, Andrew Steer, Andy Sumner, John Warburton, Jeremy Warford, Jos Wheatley, Wu Jian and Zou Ji.

And finally a shout out to the rooftop mini maze at 10 G Street NE, Washington DC, for providing moments of deep insight and inspiration.

Abbreviations and acronyms

AOSIS	Alliance of Small Island States
BASIC	Brazil, South Africa, India and China
CBA	Community Based Adaptation
CCS	CO_2 (or carbon) capture and storage
CDM	Clean Development Mechanism
CER	Certified Emissions Reductions
CH$_4$	Methane
CIF	Climate Investment Funds
CO$_2$	Carbon dioxide
COP	Conference of the Parties (to the UNFCCC)
DRR	Disaster Risk Reduction
EIT	Economy in transition
EU ETS	European Union Emissions Trading Scheme
GCF	Green Climate Fund
GCMs	Global circulation models (also referred to as global climate models)
GDP	Gross Domestic Product
GEF	Global Environment Facility
GHG	Greenhouse gas
Gt	Gigatonne (one billion tonnes)
HDI	Human Development Index
ICT	Information Communications Technology
IPCC	Intergovernmental Panel on Climate Change
LCD	Low carbon development
LDCs	Least Developed Countries

LDCF	Least Developed Countries Fund
LED	Low emissions development
MA	Millennium Ecosystem Assessment
MACC	Marginal abatement cost curve
MDB	Multi-lateral Development Bank
MDGs	Millennium Development Goals
MIC	Middle-income country
MRV	Measurement, Reporting and Verification
NAMAs	Nationally Appropriate Mitigation Actions
NAP	National Adaptation Plan
NAPA	National Adaptation Programmes of Action
NGO	Non-governmental organization
ODA	Official Development Assistance
OECD	Organisation for Economic Cooperation and Development
PPM	parts per million (of CO_2 equivalent)
RCMs	Regional climate models
REDD+	Reducing Emissions from Deforestation and forest Degradation
SCCF	Special Climate Change Fund
SIDS	Small Island Developing States
SMEs	Small and Medium-sized Enterprises
SRES	Special Report on Emissions Scenarios
SREX	IPCC Special Report on Managing the Risks of Extreme Events and Disasters to Advance Climate Change Adaptation
SSE	Steady-state economy
UN	United Nations
UNFCCC	United Nations Framework Convention on Climate Change
UNGA	United Nations General Assembly

Introduction

Climate change, the mother of all development challenges.

The mother of all development challenges

We wrote this book in the midst of the deepest global economic crisis since the Great Depression of the early 1930s. One of the victims of this 'Great Recession' was humanity's collective effort to address climate change. Before the crisis hit in 2008, a bipartisan consensus on the urgent need to tackle climate change was emerging in the United States, and similar momentum was gathering elsewhere in the world. Future generations may well look back with puzzlement at the sudden dissipation of this momentum in the face of shifting economic fortunes. While political leaders concentrated on solving the financial crisis, they lost sight of a much larger crisis.

More than twice as much heat-trapping carbon dioxide is emitted each year as a result of human activity than can be absorbed by the world's oceans and forests. As in any scientific inquiry, there are uncertainties inherent in **climate** science. Nevertheless, the evidence for human-induced climate change is overwhelming: a recent survey found that 97 per cent of the over 12,000 peer-reviewed climate science papers produced between 1991 and 2011 agree that global warming is happening and is caused by humans (Cook *et al.*, 2013).

The case for urgent action is equally compelling. Unless we rapidly reduce our **GHG** emissions, we will face a climate catastrophe that will make the Great Recession look like a walk in the park. Government bailouts can save troubled financial institutions, but there is no human power that can bail out a bankrupted planet. The impacts of climate change are likely to be more extensive, expensive and long lasting than the effects of the financial crisis. Indeed, climate change impacts may well be irreversible. We eventually bounce back from economic and financial crises, but overstepping climate thresholds may tip us into a permanently altered planetary state – a 'new normal' in which the physical and human geography of our world will be dramatically redrawn by higher sea levels, forest die-outs, and less productive and less habitable regions. The rise in atmospheric concentrations of greenhouse gases will be the critical factor shaping the future of life and human well-being on Earth.

The impacts of the climate crisis will be particularly devastating for poor people, who tend to be the most vulnerable to climate-related shocks and stresses and the least equipped to cope with them. While climate change may seem a distant concern to many in the rich world, poor people in developing countries are already struggling under the weight of weather-related disasters that climate change will only exacerbate. If unchecked, the impacts of climate change are likely to reverse decades of social and economic progress in developing countries.

Not only will climate change impact development outcomes, but the need to limit climate change will increasingly shape development choices. Collective actions to address climate change will present challenges, but also offer developing countries unprecedented opportunities, such as leapfrogging to cleaner technologies and building future-fit economies that are resilient, efficient and deliver benefits that are more equitably shared. Our planet's changing climate will define the development context for the twenty-first century.

Although the scale of the climate challenge we face is unprecedented in human history, our past holds important lessons. Recent archaeological evidence points to climatic change as a causal factor in the collapse of some ancient civilizations, and reminds us what is at stake. The collapse of the Akkadians, the Mayan, the ancient Egyptians and the Tang dynasty may all be linked to local climate disturbances (*New Scientist*, Aug 4, 2012). Unlike these doomed civilizations, which were caught off guard, we have the tremendous advantage of knowledge, foresight and advanced technology, and we may still have time to avert disaster.

We stand at a pivotal juncture in the evolution of our planet and our species. It may appear that climate protection and development goals are in direct conflict; fossil fuels are, after all, a staple of modern economies. But on closer examination it is clear that it is not the achievement of development *goals*, but the dominant *means* by which we have carried out development that is the source of conflict. Not only is there room for alternative development choices that reconcile development and climate protection, but each can succeed in the long term only if they are addressed as interconnected challenges. As the International Council for Science stated in 2010, 'Humanity has reached a point in history at which a prerequisite for development – the continued functioning of the Earth system as we know it – is at risk.' The window for action is narrow and rapidly closing. Action in the next two decades will be particularly critical.

Why this book?

As a development challenge, **climate change** resists easy categorization: it cuts across entrenched institutional divides and disciplinary boundaries. For a long time most of what we knew about climate change came from the natural and physical sciences. Although climate science generates essential insights and understanding about the behaviour of our climate system and how it is affected by human activities, it is also limited. Some of the most important questions about how to respond to climate change – such as what constitutes tolerable risk and how to determine a fair sharing of responsibility and costs across individuals, nations and generations – pertain to politics and ethics and are beyond science to answer. Sadly, politicians have mostly been unable or unwilling to make difficult decisions about how to respond to climate change, and have instead shifted the onus on to science to guide action.

This book lays a multi-disciplinary foundation for understanding the complex and tangled relationship between development, climate change, and the measures needed to minimize climate-related risks and impacts. It argues for holistic approaches that embrace insights, concepts and analytical tools from a range of individual disciplines, and transcends their divisions.

In the three decades since policy makers first became aware of climate change, they largely treated it as an environmental-pollution concern with little relevance for development choices. Climate change only started entering mainstream development discourse in the mid 2000s. In the space of a few years, climate change went from a fringe scientific and environmental concern to an issue that government leaders and development practitioners worldwide recognize as a defining challenge of our times. Developing countries' unprecedented participation in the UN climate change negotiations in Copenhagen in 2009 marked a turning point for international development as much as international environmental diplomacy.

Even as climate change begins to enter mainstream development discourse, our approaches to development and climate change remain largely segregated in practice. Climate change issues are often relegated to specialized environmental or disaster-response authorities that view them in narrow technical terms. These specialized authorities are ill-equipped to respond to the full spectrum of development challenges that climate change raises.

Like our institutions, the way our knowledge is generated also tends to be compartmentalized within specialized academic fields: climate change and development studies are rooted in different knowledge communities that often talk at cross-purposes. The continued fragmentation of our knowledge, institutions and decision-making results in climate change and development being treated as distinct and sometimes conflicting concerns.

Important cracks started appearing in the conceptual walls between climate change and development in the early 2000s. The publication of the Intergovernmental Panel on Climate Change (IPCC) Third Assessment Report in 2001 put poverty on the climate change agenda by highlighting the **vulnerability** of poor people to climate change impacts. In 2003, a multi-agency report on climate change and poverty emphasized the synergies between climate change and poverty-reduction agendas, and observed that climate change was likely to undermine achievement of the Millennium Development Goals if left unchecked (AfDB, 2003).

While there are edited volumes and individual books on various aspects of the climate–development relationship, we know of no comprehensive, accessible and multi-disciplinary treatment of the climate–development nexus. This book aims to fill that gap: it equips students and development practitioners with analytical tools and conceptual frameworks to understand and critically engage with a development landscape that is rapidly being transformed by the challenges and opportunities posed by climate change. It also aims to help readers navigate contemporary debates about alternative development pathways in a warming and resource-constrained world.

On 'development' and poverty reduction

While climate change can be scientifically defined and observed, development is value-laden and its meaning contested (Willis, 2005). It is important at the outset then to define what we mean by 'development'. Sumner and Tribe (2005) identify three broad definitions:

- The process of structural change as observed in demographic, societal and economic shifts (**development as process**). The advent of agriculture, emergence of cities and nation-states, and the industrial revolution are all key turning points in humankind's development. Defining characteristics of development today include urbanization, industrial development, modernization and consumerism.
- Deliberate efforts aimed at improving human well-being (**development as project**). Development in this sense refers to a realm of human endeavour, manifested in professions and organizations and driven by goals, strategies, policies, plans and projects.
- A dominant discourse of social progress (**development as discourse**). Development in this sense has been dominated by the Enlightenment tradition, and its interpretation of history as the progressive, linear unfolding of universalizing reason via science, technology and rational institutions (also known as 'modernization').

Since this book is chiefly concerned with development *choices* and *outcomes* – and the manner in which they can conflict with or assist climate protection – the term 'development' is employed in the second sense, referring to efforts to improve human well-being, with a focus on poverty reduction. National actors have led these efforts, with support from rich countries in the form of development aid. At the international level, the Millennium Development Goals (see Box 0.1) provide a common framework to guide development efforts till 2015. A new development framework is being negotiated for the post-2015 period as this book goes to print.

Poverty is commonly associated with low levels of income, but it is more accurately defined as the opposite of well-being. In the narrowest sense, well-being is equated with the satisfaction of material and physiological needs and wants, and is tantamount to being well-off, financially or materially. This reductionist conception of well-being is adopted in mainstream economics, where well-being is reduced to the money metric: material deprivation is thus reflected as a lack of income. Development and poverty reduction in this narrow economic conception is focused on increasing peoples' incomes,

Box 0.1

The Millennium Development Goals

In 2000, leaders of 189 countries met at the United Nations in New York and endorsed the Millennium Declaration, a commitment to work together to build a safer, more prosperous and equitable world. The Declaration was translated into a roadmap setting out eight time-bound and measurable goals to be reached by 2015, known as the Millennium Development Goals, namely:

1   Eradicate extreme poverty and hunger

2   Achieve universal primary education

3   Promote gender equality and empower women

4   Reduce child mortality

5   Improve maternal health

6   Combat HIV/AIDS, malaria and other diseases

7   Ensure environmental sustainability

8   Develop a global partnership for development

Source: www.un.org/millenniumgoals

which is best achieved in the aggregate through increasing a country's per capita gross domestic product (GDP).

Amartya Sen champions a more sophisticated perspective on poverty and well-being, which has gained widespread acceptance. In his view, the importance of a person's freedoms and capabilities is paramount. For Sen, well-being depends not simply on material conditions, but on an individual's ability to lead a meaningful life that he or she has chosen and has reason to value. Individuals are thus socially constituted creatures that derive meaning and fulfilment from participation in society, and are driven by complex motives beyond the satisfaction of material desires. In this perspective, popularized in Sen's book *Development as Freedom*, the enlargement of peoples' opportunity sets, or freedoms, is both the end and the means of development (Sen, 1999). We use Sen's broad conception of well-being in this book.

Section One: Understanding the building blocks

Although this book focuses on the implications of climate change for human development, a rudimentary knowledge of climate science is crucial. In Section One, we lay down the building blocks for understanding and framing the nexus between climate change and development.

Chapter 1 walks the reader through the basics of the climate system, climate change to date, projections of future climate change, and some of the potential impacts on developing countries. We draw heavily on the assessment reports of the IPCC as summaries of robust, peer-reviewed, international scientific literature. The attempts by climate sceptics to discount human-caused climate change are regarded as scientifically weak, but significant in their impact on opinions and behaviour around the world. Readers can visit counter-sceptic websites such as www.skepticalscience.com and www.realclimate.org to learn more about these debates.

In **Chapter 2**, we introduce the 'climate–development nexus', that is, the complex interactions between climate, environmental and socio-economic systems that underpin the linked challenges of climate protection and development. We argue that actions in the climate and development spheres share the common normative underpinnings of **equity, social justice** and sustainability.

Responding to climate change requires that we deal with both its causes and its consequences. Climate change **mitigation** addresses the causes of climate change, while **adaptation** focuses on reducing vulnerability to the climate change impacts that cannot be avoided through mitigation. Both mitigation and adaptation responses are fundamentally linked to development processes. For example, expanding access to modern energy is a development priority. But if not handled with an eye towards climate impact, expanded energy access could increase **greenhouse gas** emissions. Similarly, the ability to adapt to climate change depends on the essential ingredients of development, including equity, information, effective institutions and voice in decision-making, as well as income levels. Chapter 2 explores the concepts of vulnerability and adaptation in the context of development before considering the integration of approaches to both mitigation and adaptation in development.

Section Two: Responding to climate change as a development challenge

This section of the book responds to the explosion of thinking, research, policy and action linking climate change and development in four response areas: international rule-setting, mitigation, adaptation and finance. Rather than simply describe climate responses to date, Section Two offers analytical tools for considering each response area. These analytical tools will enable readers to ask smart questions about the climate change and development nexus as policy and action evolve in the coming years.

Chapter 3 introduces the international climate policy arena that the United Nations Framework Convention on Climate Change (**UNFCCC**) has dominated since the Rio Earth Summit in 1992. We examine the UNFCCC's basic architecture and evolution, with a focus on the developing-country context. We then set out key concerns that will define future responses to climate change, including equity and justice in sharing the burden of emission reduction, and the transfer of finance and technology from richer to poorer nations.

In **Chapter 4**, we examine mitigations responses. We start by considering the historical link between rising income, rising energy use and rising greenhouse gas emissions. Development will inevitably entail increased emissions in most poor countries, particularly through meeting energy supply needs. Chapter 4 presents the different analytical approaches to understanding the options for lower emissions development. It goes on to outline the main policy options for mitigation, examining the sectoral contexts pertinent to developing countries.

Chapter 5 examines the process of adaptation and its links to development. Although mitigation has driven international policy agendas, adaptation issues have emerged in response to the concern of poorer nations and development organizations rooted in the recognition that poor people are particularly vulnerable to climate-related shocks and stresses. Chapter 5 charts the evolution of different analytical approaches to adaptation. Although these approaches are often used in combination, they understand, analyse and tackle the impacts of climate change differently.

Chapter 6 considers some of the key issues around climate finance, an issue that will be crucial to any future global deal for tackling climate change. After examining the definition, rationale and requirements for climate finance, the chapter introduces some of the existing sources of international climate finance available to developing countries. Acknowledging the dynamic nature of climate policy and finance, the chapter introduces a set of key analytical themes for examining existing and new funding schemes as they emerge, including governance, sources and spending. Notably, most international attention to date has focused on governance and sources, with far less emphasis on where and how to spend finance in ways that can simultaneously alleviate poverty and tackle climate change.

Section Three: Development futures in a transformed world

The last three chapters of the book, contained in Section Three, are forward looking and focus on why and how development must be re-framed to deliver more equitable and sustainable outcomes. It sets out different critiques of 'development as usual' and explores competing visions of development in a warming and resource-constrained world.

Chapter 7 considers the changing context of development, shaped by the interlocking challenges of widening inequalities, growing natural-resource scarcities, deteriorating ecosystems and shifting power relations. This scene-setting chapter will help to ground the debate about alternative modes of development.

The definition and construction of a problem determines how it is addressed. In **Chapter 8**, we consider alternatives to GDP-centric paradigms of development. These include alternatives that are consistent with low-emissions and climate resilient development, and contribute to social justice and poverty reduction. Chapter 8 also sets out the basic conditions and first steps that need to be taken on in the transition towards climate-smart development.

Chapter 9 is a concluding chapter that reflects on the current status of the science and actions taken to date, and sets out five critical challenges that will determine progress in the decade ahead.

Student briefing

This book offers a primer on the challenges, responses and alternative pathways to tackling development in a changing climate. It draws together insights and perspectives from natural and social sciences, and from local, national and global analyses. Focusing on the concerns and perspectives of poor countries and poor people, the book moves the reader from the basics of the climate change and development nexus, through the potential responses in main-issue areas, to an examination of climate change and development as central to the sustainability of human progess. It posits the need for alternative paradigms to reconcile climate change and development.

This book can serve as a jumping-off point for further investigation. At the end of each chapter, we provide links to suggested reading and useful websites. The lists of suggested readings focus on journal articles and books, but given the rapidly changing nature of climate and development policy and practice, readers should also consult 'grey literature' such as reports and policy briefings. Internet search engines such as Google, Google Scholar and Google Books are one way to find these resources. Another is to read websites that summarize and link to full reports, such as the long-running Eldis resource guide on climate change (www.eldis.org/go/topics/resource-guides/climate-change). List-serves, where members post announcements of publications and events, are another useful resource. The most well-used climate change-related list-serve is probably Climate-L, which is linked to the Climate Change Policy and Practice site at the International Institute for Sustainable Development (IISD) in Canada (http://climate-l.iisd.org).

Many of the topics addressed in this book are rapidly evolving. Readers should therefore refer to this book's companion website www.climateandevelopment.com for updates on latest developments and links to key knowledge resources on the climate-development nexus. A glossary is provided at the end of this book for key terms and concepts, indicated in the text in bold. For a more complete reference guide we refer you to the *The Complete Guide to Climate Change* ((2009), Abingdon: Routledge), by Matthew Spannagle and Brian Dawson.

A book of this length cannot do justice to the breadth and depth of issues caught up in the climate change and development nexus, but it can guide further exploration. We hope that this book will inspire and energise a range of people – from students and community groups, to development practitioners and policy makers – to take action to meet the mother of all development challenges.

SECTION ONE
Understanding the building blocks

1 Science, drivers and impacts of climate change

The science of climate denial.

- Climate science and climate scepticism
- Climate and the climate system
- What drives the enhanced greenhouse effect?
- Defining 'climate change'
- Recent climate change and future projections
- Regional impacts of climate change in the developing world
- From climate science to policy response

Introduction

This book focuses on the implications of **climate change** for **human development** in developing countries. In recent years, climate change approaches that do not rely as heavily on scientific models to project future change have become more prevalent, especially in developing-country contexts, where scientific research is often less advanced. Despite the growing range of disciplinary perspectives on climate change and development, **climate** science retains a privileged position in informing debates and action. Consequently, a rudimentary knowledge of atmospheric systems and climate science is essential for understanding the interactions between climate change and development. It is important also to recognize that scientific findings are inherently contingent and science is ever evolving: it does not provide settled and permanent truths. Instead, it puts forward tentative insights and hypotheses about the workings of our physical world, which are iteratively tested and superseded or improved.

We begin this chapter by introducing the basic functioning of the global climate system and how human impact on the system has generated climate change. We then describe what climate science tells us about climate change, including recent historical change, modeling of future climate change and potential regional impacts. Finally, we consider the mix of economics, politics and value judgements that determine our responses, both to mitigate the processes of global climate change and to adapt to impacts that cannot be avoided.

1. Climate science and climate scepticism

Our understanding and responses to climate change are framed predominantly by the methods and findings of climate science. Climate science draws largely from the physical sciences, particularly atmospheric physics and chemistry. Thus a scientific method of understanding physical and human processes is at the heart of what we know about, and how we act upon, climate change. This positivist method formulates, tests, and modifies hypotheses through systematic observation, measurement, and experiment. However, as we explore in the final section of this chapter, climate science is not always treated objectively; it can be used to support different arguments and interests. The media and other thought leaders may pay special attention to minority views that challenge scientific consensus. The interplay of politics and science is therefore crucial to understanding decision-making responses to climate change (Pielke, 2010; Dessler, 2012).

This chapter draws heavily from the Assessment Reports and Special Reports of the Intergovernmental Panel on Climate Change (IPCC) (see Box 1.1). We encourage readers to familiarize themselves with

Box 1.1

The Intergovernmental Panel on Climate Change

The IPCC was initiated in 1988 as a joint effort of the World Meteorological Organization and the UN Environment Programme. The IPCC is made up of thousands of international scientists who review scientific, technical and socio-economic information from across the world to provide periodic assessments of the state of climate change, its potential consequences and possible responses. IPCC reports are policy-relevant but not policy-descriptive. National governments endorse its findings and it is widely regarded as the authoritative scientific body on climate change.

The IPCC has completed Assessment Reports in 1990, 1995, 2001, and 2007, with the Fifth Assessment Report released across 2013–2014. Readers are strongly encouraged to read the latest findings as they are published. It has expanded its focus over time, adding new and emerging focuses to its ongoing work; issues of equity, alternative development pathways and sustainable development have been given greater prominence in the more recent reports.

The IPCC assessments are divided into three working groups covering:

- the physical science basis of climate change

- climate change impacts, adaptation and vulnerability

- mitigation of climate change

IPCC bodies also produce methodological guidance, technical papers and special reports on specific topics, including: extreme events and disasters, aviation, regional impacts of climate change, technology transfer, emissions scenarios, land use, land use change and forestry, carbon dioxide capture and storage, the ozone layer, and renewable energy.

these reports, especially the more up to date Fifth Assessment Report published in 2013–2014. The overwhelming weight of scientific opinion supports the main findings of the IPCC assessments, which, because they are based on consensus and draw on a wide range of scientific studies, are likely to present conservative findings. Nevertheless, we acknowledge the importance of the counter-claims of climate sceptics in so far as they influence behavioral and policy responses to climate change, even though they generally rely on a scientifically weak or partial evidence base.

The large majority of climate scientists agree that human-caused climate change is real. Analysis of climate researchers and their publication and citation data by Anderegg *et al.* (2010) shows that more than 97 per cent of the most published climate researchers support the tenets of human-caused climate change outlined by the IPCC. Their analysis also shows that the relative climate expertise and scientific prominence of the researchers who doubt these tenets are substantially below the researchers who agree with the tenets. Despite the fact that not all climate researchers are equal in scientific credibility and climate expertise, the media tends to present both sides in climate change debates, which can contribute to ongoing public misunderstanding of the scientific evidence.

Good scientists are 'sceptical' in the sense that they question and test established theories. However, 'climate scepticism' and 'climate denial' are terms generally used to describe those who deny climate change and/or its links to human interference. Climate scepticism is underpinned by broader environmental scepticism that:

- denies the seriousness of environmental problems;
- denies the credibility of the evidence for such problems;
- questions the importance of environmentally protective policies (Jacques *et al.*, 2008: p. 354).

Climate scepticism and doubt about climate science, which by its nature contains some inherent uncertainty, are often espoused by people wishing to preserve the status quo. Many of the sceptics' counter-claims can be traced directly to groups that would suffer from changes to the status quo. These entities include oil companies, energy-intensive industries, and countries with economies dominated by fossil fuel extraction or forestry. Individual consumers who are reluctant to change their own behavior may also draw on climate scepticism to justify their actions (or inaction).

The 'climategate' episode of 2009, when private emails by climate scientists at the University of East Anglia were released on the internet, brought climate scepticism to the fore (Nerlich, 2010). Critics honed in on the use of the mathematical term 'trick' in adjusting data and discussions over whether to accept for publication work they considered flawed. In 2010 an independent panel cleared the University of any scientific impropriety or dishonesty, but some vocal bloggers presented the emails as evidence that climate scientists were colluding to fix data and keep contrary scientific views out of peer-reviewed journals. With the email leak timed just before the Copenhagen international climate negotiations, questions about the validity of the scientific process underpinning the meeting simultaneously weakened the negotiations and undermined public concern about the climate change challenge.

Countering climate denial, Simon Dietz and colleagues (2007: 250) rebut sceptics, arguing,

> Those who deny the importance of strong and early action should explicitly propose at least one of three arguments: (i) there are no serious risks; (ii) we can adapt successfully to whatever comes our way, however big the changes; (iii) the future is of little importance. The first is absurd, the second reckless and the third unethical.

There are numerous websites dedicated to the latest climate science (e.g., www.realclimate.org) and rebutting climate sceptical views (e.g., www.skepticalscience.com). We encourage readers to become acquainted with these sites and the core arguments/rebuttals.

There are also more scientific critiques of the dominant paradigm of climate science. Drawing on insights from Gaia theory, which he helped pioneer, James Lovelock (2010) argues that prevailing scientific methods fail to treat the climate and wider earth system as dynamic and self-regulating. Rather than responding in a steady linear way, the system has **tipping points** or thresholds at which the system's behavior abruptly changes and the system gravitates towards a new equilibrium, a process of self-regulation which is out of our control. Lovelock has also been critical of the over-reliance on mathematical climate models, which assume the earth responds linearly and passively to change. Such models, he argues, rely on a limited empirical understanding of the interactions of different parts of the climate system, especially the oceans and the responses of ecosystems to climate change. He calls for a new scientific approach that models the Earth as a single physiological system rather than divided into separate parts for study by different disciplines.

By contrast, recent studies of how model projections in the past have measured up against the reality of what has actually happened suggest that climate models have been surprisingly accurate (Allen *et al.*, 2013). This book considers the IPCC process of international assessment based on peer-reviewed literature a robust scientific basis for policy and decision making. Uncertainty is a defining feature of science, and hypotheses/theories are provisional until disproven by robust evidence. This does not mean that the scientific method is flawed:

> In the real world, it is impossible to prove that scientific theories are right in every circumstance; we can only prove that they are wrong. This provisionality can cause people to lose faith [or patience] in the conclusions of science, but it shouldn't. The recent history of science is not one of well-established theories being proven wrong. Rather, it is of theories being gradually refined. Newton's laws of gravity may have been superseded, but they are still accurate enough to be used for many purposes.
>
> (*New Scientist*, 2011)

2. Climate and the climate system

Climate is usually defined as the average and variability of temperature, precipitation and wind over a period of time, ranging from months to millions of years (IPCC, 2007). Thirty years is the most commonly used defining period. Climate is related to but different from weather – the shorter-term, day-to-day manifestation of these phenomena. Whereas weather is notoriously difficult to predict, climate is a long-term average over time, and hence variations in the future climate are easier to predict than individual weather events.

The common misperception that a particularly cold winter is evidence against global warming is attributable to the confusion between weather and climate. Scientists acknowledge that there are always hot and cold extremes, but when weather is averaged over space and time, the data show that the globe has been warming in recent decades. Importantly, climate change leads to changes in the frequency, intensity, spatial extent, duration, and timing of weather events, including extreme weather, and can result in unprecedented extreme weather and climate events (IPCC, 2012). As the number of unusual weather events grows, they can be statistically linked with increasing confidence to a changing climate.

The Earth's climate depends on the balance between the energy received from the sun as solar radiation, the energy radiated back from the Earth into space, and the energy exchanged between parts of the climate system. The climate system comprises the atmosphere, land, oceans, ice and living things on Earth, all of which interact. The climate system changes over time through both its internal interactions and external factors known as **forcings**. Forcings include natural phenomena such as volcanic eruptions and variations in the sun, and human-caused changes in the atmosphere and the reflectivity of the Earth and atmosphere. The overall global climate balance can therefore alter as a result of changes in:

1 incoming solar radiation (such as changes to the Earth's orbit or the sun's activity);
2 outgoing radiation from Earth back towards space (such as by changing **greenhouse gas** concentrations);
3 the amount of solar radiation that is reflected, called 'albedo' (such as changes in cloud cover, atmospheric particles or land surface cover).

Total outgoing radiation is related to the presence of certain natural greenhouse gases in the atmosphere, including water vapor (H_2O), carbon dioxide (CO_2), methane (CH_4), ozone (O_3), and nitrous oxide (N_2O). These gases act as a partial blanket, trapping outgoing radiation from escaping back into space. This 'natural **greenhouse effect**' enables the Earth to maintain a temperature where most water exists in liquid form and human life is possible. The increase in this warming effect caused by human activity (through increases in the amounts of greenhouse gases and small particles) is known as the *enhanced greenhouse effect*.

The concentration of greenhouse gases in the atmosphere is affected by interactions between components of the climate system. For CO_2, one of the most significant greenhouse gases, this interaction is governed by the carbon cycle, which enables carbon to be recycled and reused throughout the climate system. Plants absorb CO_2 from the atmosphere and convert it (along with water) into carbohydrates via photosynthesis. In the industrial era, human activities have added greenhouse gases to the atmosphere, primarily by burning fossil fuels and clearing forests.

Changes to the composition of the atmosphere can drive climate forcings, but feedback mechanisms in the climate system can have a significant effect in either amplifying or diminishing the effects of any change. For example, snow and ice melting at high latitudes resulting from an enhanced greenhouse effect reveals darker soil and water surfaces. This alters the third element of the global climate balance noted above (albedo), as darker surfaces reflect less and absorb more of the Sun's heat, creating a positive feedback of further warming and melting (IPCC, 2007). These feedback effects help determine the eventual impacts of changes driven by other climate forcings, including changes in the atmospheric concentration of greenhouse gases.

Our understanding of the climate system and its processes has evolved over many years, with large leaps in recent decades, particularly aided by modern computers that can simulate sophisticated climate system models (see Box 1.2).

Box 1.2

A potted history of climate science

Building on Joseph Fourier's calculations of the atmosphere's transmission of radiation, John Tyndall's 1864 experiments first demonstrated the differential absorption of infrared radiation by different gases, highlighting the strong absorptive capacity of water vapour, hydrocarbons and carbon dioxide. In 1896, Swedish scientist Svante Arrhenius was the first to estimate the alterations to the climate system caused by human activity, in part as a contribution to debates around the causes of ice ages. Noting that a cooler or hotter atmosphere would create enhanced effects through positive feedbacks caused by water vapour and changes in surface albedo, he calculated that doubling atmospheric CO_2 would lead to a total warming of 5–6°C, while halving it would produce an ice age. This 'greenhouse effect' was only one of many theories about climate change at the time, and even Arrhenius assumed that such changes would take place over tens of thousands of years.

By the 1950s and 1960s, scientists began to investigate the claims of amateur scientist G.S. Callendar, who believed greenhouse warming was impending and that it could occur over centuries rather than millennia. The rise of the global environment movement in the 1970s increased consciousness, but research into the role of dust and smog particles in blocking sunlight and cooling the world, along with evidence of a cooling trend in the Northern Hemisphere, diverted policy makers. Climate research was revealing the complexity of the system and its sensitivity to external influences such as volcanic eruptions and solar variations, as well as human activity. After the summer of 1988 measured as the (then) hottest on record, corporations and individuals opposed to government regulation began commissioning and circulating climate counter-claim reports.

By the 1990s, improved models could reproduce detailed patterns of the temperature and rainfall changes actually observed in different world regions over the past century. Models predicted the temporary cooling from a huge volcanic explosion in 1991, and showed significant warming when greenhouse gases were added. Studies of Antarctic ice cores verified the link between carbon dioxide and temperature used in computer models: historically, a doubling of carbon dioxide coincided with a roughly 3°C temperature rise.

The IPCC's Third Assessment Report of 2001 completed the 'discovery' of global warming. The less probable but more severe effects of climate change also became apparent. These include the dangerous changes in ocean circulation, the effects of collapsing ice sheets on sea-level rise caused by thermal expansion of the oceans, and positive feedbacks whereby ongoing warming was itself leading to changes that would generate still more warming.

In 2007, the Fourth IPCC Assessment Report reported with greater confidence that humans were changing the climate. It noted that the impacts of global warming – more deadly heat waves, stronger floods and droughts, heat-related changes in the ranges and behaviour of sensitive species – were already visible in some regions. The 2013 Fifth Assessment Report increased this confidence level even further, stating that it is extremely likely that human influence has been the dominant cause of the observed warming since the mid-20th century.

Source: Adapted from Weart, 2008; IPCC, 2013.

3. What drives the enhanced greenhouse effect?

Throughout Earth's history, its climate has changed in response to natural variations in climate forcings. Contemporary climate change is fueled primarily by human influence on the climate system through the introduction of pollutants. These pollutants can either warm or cool the Earth, and can be short-lived (persisting for days to weeks) or long-lived (persisting for decades to centuries). Sulphate aerosols from burning coal are short-lived pollutants, and the four main human-caused ('anthropogenic') greenhouse gases are long-lived.

While greenhouse gases exist in the atmosphere naturally, human activity, particularly since the start of the industrial age in roughly 1850, has increased the concentrations of four gases:

- Carbon dioxide (CO_2), which has increased principally due to greater fossil fuel use since the industrial revolution and from deforestation, which releases CO_2 and also reduces the number of plants available to take up CO_2.

- Methane (CH_4), which has risen due to greater agricultural activities, landfill waste and use of natural gas as fuel.
- Nitrous oxide (N_2O), which is emitted by human activities such as fertilizer use and fossil fuel burning.
- Halocarbon gases, principally the chlorofluorocarbons (CFCs), which are used as refrigeration agents and in other industrial processes. The emission of these gases has decreased dramatically since they were discovered to deplete stratospheric ozone.

Other greenhouse gases include:

- Ozone (O_3), a naturally occurring greenhouse gas, has increased through chemical reactions with other gases released by human activity.
- Water vapor (H_2O) is the most abundant and important greenhouse gas in controlling the greenhouse effect. Human activities have only a small direct influence on the amount of H_2O, notably through methane emissions. However, by changing the climate itself, humans can affect water vapor, as a warmer atmosphere will contain more water vapor.

Greenhouse gases increase the warming effect by absorbing outgoing infrared radiation in the atmosphere, with increases in CO_2 causing the largest forcing since 1750. This helps explain why the climate change debate has largely centered on carbon dioxide emissions and their management. However, the influence of other effects not directly related to greenhouse gases should not be discounted. Indeed, some commentators argue that the focus on CO_2 limits the impact of policy responses to climate change:

> Make no mistake: carbon dioxide matters a great deal. However, a key implication of recognizing the diversity of human influences on the climate system is that even if we were to meet the challenge of stabilizing concentrations of carbon dioxide (and even other greenhouse gases) in the atmosphere at a low level, we would not have solved the larger challenge of addressing human influences on the climate system, because we have so many other influences.
>
> (Pielke, 2010: p. 18)

Those other important human influences on the climate system include changes in land use and land cover such as pastures, croplands, forests and snow and ice, which alter the reflective property of the land surface (including the effect of soot which darkens snow surfaces). Aircraft condensation trails, which reflect solar radiation and absorb infrared radiation, also influence the climate system.

Aerosols – small particles in the atmosphere – exert important radiative forcings and are generally linked to a cooling effect on the global climate system. Derived both from naturally occurring sources (such as volcanic eruptions) and from human activities (such as burning fossil fuels), aerosols directly affect the reflection and absorption of solar and infrared radiation in the atmosphere and indirectly by affecting cloud properties. Temporary negative radiative forcings commonly follow major volcanic eruptions for two or three years. Natural cycles in the sun's radiation also exert a radiative forcing effect. Crucially, however, the radiative forcing from human activities exerts a much greater influence on the climate than the estimated radiative forcing from changes in natural processes (IPCC, 2007).

Although emissions of anthropogenic (human-caused) greenhouse gases have grown significantly since the start of the industrial age, growth has accelerated in recent decades: between 1970 and 2004, anthropogenic emissions grew by 70 per cent (see Figure 1.1). Energy supply now accounts for over a quarter of emissions, while forestry, agriculture, transport and industry each account for between 13 per cent and 20 per cent. From a development perspective, these emissions are spread unevenly across different parts of the world, reflecting regional difference in economic development, energy mix, land cover and land cover change (see Figure 1.2). The richer industrialized countries (the Annex 1 group in the UN Climate Change Convention) have the highest emissions per head of population, but emissions per unit of economic output (GDP adjusted for purchasing power parity) are lower than developing countries, indicating the potential for enhancing emissions efficiency in many countries.

Determining the response required to limit the enhanced greenhouse effect (known as climate change **mitigation**) requires understanding the sensitivity of the climate system to different levels and mixes of atmospheric greenhouse gas concentrations, as well as the effect of other human activities on radiative forcing, including aerosols and changes to surface reflectivity (albedo). The higher the level at which atmospheric greenhouse gas concentrations are stabilized, the higher the expected increase in global average temperature, which serves as a common proxy for the wider range of climate change impacts.

Our response must also account for the long-term 'lock-in' to climate impacts due to the lag times between emissions and impacts (and therefore between reducing emissions and reducing their impacts) (IPCC, 2001a). This lag results from the slow breakdown of many gases in the atmosphere and the inertia of other elements of the climate system. The result is that even if total greenhouse gas

Figure 1.1 *Global anthropogenic GHG emissions: (a) Global annual emissions of anthropogenic GHGs from 1970 to 2004; (b) Share of different anthropogenic GHGs in total emissions in 2004 in terms of carbon dioxide equivalents (CO$_2$-eq); (c) Share of different sectors in total anthropogenic GHG emissions in 2004 in terms of CO$_2$-eq.*

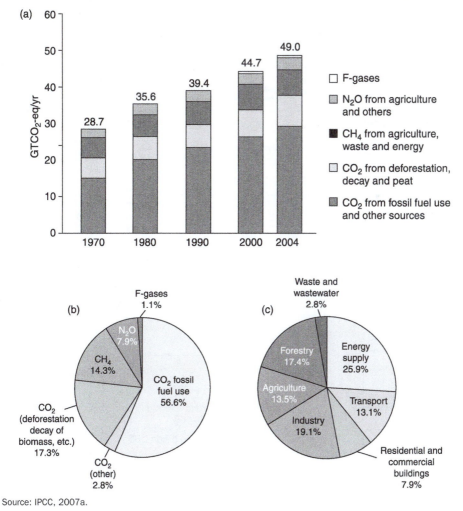

Source: IPCC, 2007a.

Note: forestry includes deforestation.

Figure 1.2 *Regional distribution of GHG emissions by population and by GDPppp:*
(a) Distribution of regional per capita GHG emissions according to the population of
different country groupings in 2004 (see Chapter 3 for definitions of country groupings);
(b) Distribution of regional GHG emissions per US$ adjusted for purchasing power parity, or
GDP_{PPP} of different country groups.

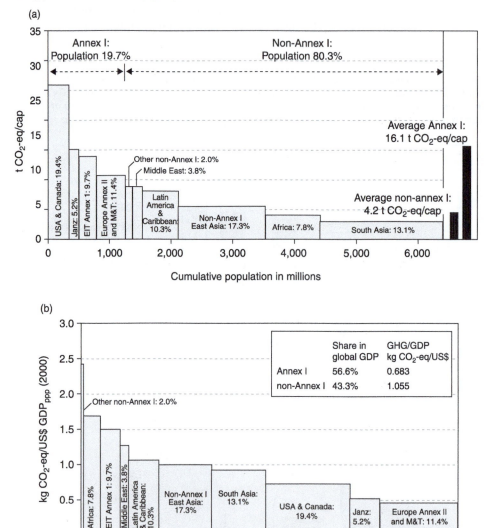

Source: IPCC, 2007a.

Note: the percentages in the bars in both panels indicate a region's share in global GHG emissions.

emissions peak in the next decade, their atmospheric concentrations will only stabilize over a number of centuries, as will temperatures, while sea level rise effects may continue over several millennia. Conversely, human activity that causes cooling effects, such as aerosols from burning coal, have a much shorter lag-time because they are removed relatively quickly from the atmosphere through wind and rain. As a consequence, reductions in aerosols associated with urban clean-air initiatives or using less coal-fired energy sources may reduce global warming in the long term but increase the warming effect from incoming radiation in the short term as the cooling effect of aerosols is reduced (Andreae *et al.*, 2005).

4. Defining climate change

The distinction between natural and human-caused forcings is reflected in the different definitions of climate change. Each definition points climate policy in a particular direction, which in turn influences the institutions dedicated to climate change and drives climate science to answer questions pertinent to that particular framing of the problem.

The IPCC defines climate change as a 'statistically significant variation in either the mean state of the climate or in its variability, persisting for an extended period (typically decades or longer)' (IPCC, 2001b: p. 711). The IPCC notes that climate change may be attributable to either natural internal processes or external forcings, or to persistent anthropogenic changes in the composition of the atmosphere or in land use.

In contrast, the UN Framework Convention on Climate Change (**UNFCCC**), the international body for tackling climate change (see Chapter 3), defines climate change solely as a human-caused phenomenon: 'a change of climate which is attributed directly or indirectly to human activity that alters the composition of the global atmosphere and which is in addition to natural climate variability observed over comparable time periods'. (UNFCCC, 1992: Article 1.2). This definition restricts climate change to the changes that can be attributed to human activities, focusing on **GHG** emissions from human activity.

In this book, we adopt the IPCC definition on the grounds that: (i) it is the more scientifically appropriate definition, and (ii) it provides a sounder basis for intervention to address the impacts of climate change in a development context. The IPCC definition also helps us to

Figure 1.3 *The effect of changes in temperature distribution on extremes. Different changes in temperature distributions between present and future climate and their effects on extreme values of the distributions: (a) effects of a simple shift of the entire distribution toward a warmer climate; (b) effects of an increase in temperature variability with no shift in the mean; (c) effects of an altered shape of the distribution, in this example a change in asymmetry toward the hotter part of the distribution.*

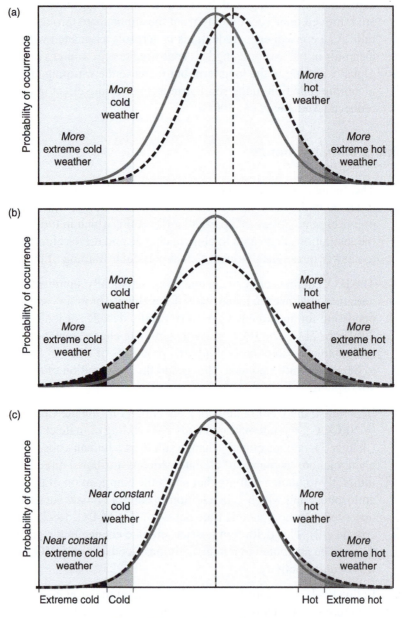

Source: IPCC, 2012.

consider climate change in terms of both changes in the average ('mean') state of the climate and changes related to climate variability (the distribution of values around that mean).

This is illustrated in Figure 1.3: in the top diagram, an increase in the mean shifts the entire distribution toward a warmer climate, eliminating very cold winter conditions and increasing extremely hot conditions. For example, an absence of cold winters has been linked to the proliferation of bark beetle and subsequent widespread tree deaths in the southwestern United States; increasing heat extremes have also been linked to the rise in forest fires. The second diagram in Figure 1.3 shows increased temperature variability with no shift in the mean. This increases the greater extremes of both hot and cold weather. The third diagram illustrates both of these changes happening at once, changing the average conditions but also the distribution of hot and cold weather around that average (IPCC, 2012).

This distinction between changes in mean and variability is important. Climate model outputs and projections are often presented in terms of changes to the mean of a climate-related unit of measurement over time (for example, a 1°C rise in temperature, or 5 per cent decrease in rainfall by 2050). In reality the variation around this average may be the most important determinant of climate-related impacts and **adaptation** to impacts. For example, for poor farmers, the average rainfall change may matter much less than changes to the number of extremely dry or extremely wet days in the year, or changes in the distribution of rainfall across the seasons. Global mean changes also mask the significant regional variations in climate change. There is also inherent uncertainty in model outputs and projections, which is communicated through assessments of the likelihood of occurrence and robustness of the evidence (see Box 1.3).

Box 1.3

Communicating scientific uncertainty: The IPCC approach

As a predictive science of a complex system, climate change models outputs contain multiple sources of uncertainty. To communicate this uncertainty figures are often expressed as an average number within a range (for example 1.8°C [1.3 to 2.3]). Where uncertainty in specific outcomes is assessed using expert judgement and statistical analysis of a body of different sources of evidence (e.g. observations or model results), the IPCC uses likelihood ranges to express the assessed probability of occurrence:

- virtually certain >99%;

- extremely likely >95%;

- very likely >90%;

- likely >66%;

- more likely than not > 50%;

- about as likely as not 33% to 66%;

- unlikely <33%;

- very unlikely <10%;

- extremely unlikely <5%;

- exceptionally unlikely <1%.

Where uncertainty is assessed qualitatively, the confidence level is based on the level of agreement between different sources of evidence as well as the amount and quality of evidence. Evidence is regarded as robust where there are multiple, consistent and independent lines of high-quality evidence.

Source: Mastrandrea et al., 2010; IPCC, 2012.

5. Recent climate change and future projections

Along with historical data, climate models underpin our understanding of the changing climate system. Global circulation models (GCMs, also referred to as global climate models) calculate mathematical representations of the different components of the climate system and their respective interactions, largely based on the laws of physics. The ability of these models to represent the climate system accurately is tested by comparing the model simulation of the past climate with historical data, as well as by comparing different models. The models' abilities to simulate both historical climates and climate changes and short-term weather prediction and seasonal forecasting indicate that they can accurately represent future climates and climate changes.

GCMs project future climate change according to a variety of global greenhouse gas emissions scenarios (known as 'SRES') based on storylines for how human activity over the next 100 years will affect emissions over the next 100 years (see Box 1.4). The results of GCMs can also feed regional climate models (RCMs), which provide greater geographical detail of climatic parameters. Climate change scenario information can then be fed into climate impact models and **vulnerability** assessments that link the projected climate scenarios with their effects on different human and ecological systems.

The progression of the four IPCC Assessment Reports since 1995 reflects our improved understanding of climate science and human influence on the greenhouse effect. In 1995, the Second Assessment Report concluded that 'the balance of evidence suggests a discernible human influence on global climate' (IPCC, 1995a: p. 22). The Third Assessment Report strengthened that finding, noting that 'there is new and stronger evidence that most of the warming observed over the last

---

Box 1.4

Future emissions scenarios

Climate models use sets of different plausible future scenarios of greenhouse gas emissions to derive projections of climate change under different mitigation pathways in the future. Previously, these were based on 'SRES' scenarios – a set of six potential future storylines that depend on different levels of population growth, economic growth, trade and technology that in turn influence greenhouse gas emissions.

The three SRES A1 storylines assume a world of very rapid economic growth, a global population that peaks in mid-century and rapid introduction of new and more efficient technologies. B1 describes a convergent world with more rapid changes in economic structures toward a service and information economy. B2 describes a world with intermediate population and economic growth, emphasizing local solutions to economic, social, and environmental sustainability. A2 describes a very heterogeneous (diverse) world with high population growth, slow economic development and slow technological change.

The IPCC Fifth Assessment Report revised SRES to define a set of four new scenarios called Representative Concentration Pathways (RCPs), which are based on different levels of radiative forcing in year 2100 relative to 1750. These four RCPs include one mitigation scenario leading to a very low forcing level (RCP2.6), two stabilization scenarios (RCP4.5 and RCP6), and one scenario with very high greenhouse gas emissions (RCP8.5).

Source: IPCC, 2007; IPCC, 2014.

---

50 years is attributable to human activities' (IPCC, 2001a). The Fourth Assessment Report states that the warming of the global climate system is now 'unequivocal' based on evidence from increases in global

average air and ocean temperatures, widespread melting of snow and ice and rising global average sea level (IPCC, 2007a: p. 30). The Fifth Assessment Report (IPCC, 2014) increased the confidence level even further than humans have caused climate change, stating that it is 'extremely likely' that human influence has been the dominant cause of the observed warming since the mid-20th century.

Global surface temperatures are at least 0.74°C above pre-industrial levels, with increases greater at higher northern latitudes; Arctic temperatures have increased at almost twice the global average rate in the past 100 years (IPCC, 2007a). Historic observations of sea level rise and decreases in snow and ice extent are also consistent with the warming trend (see Figure 1.4). Scientific confidence is high that cold days, cold nights and frosts have become less frequent over most land areas in the past 50 years, while hot days and hot nights have become more frequent (IPCC, 2012). Precipitation trends from 1900 to 2005 showed significant increases in eastern parts of North and South America, northern Europe and northern and central Asia whereas precipitation declined in the Sahel, the Mediterranean, southern Africa and parts of southern Asia (IPCC, 2007a).

Importantly, the observed temperature rise to date has been reduced because of the cooling effect of reflective aerosols and the decades it takes for ocean temperatures to warm up following increased trapping of infrared radiation (Andreae et al., 2005; IPCC, 2007a). These effects account for a further 1°C of temperature rise that may be realized in the future, meaning that even at current levels of forcing, we are committed to nearly 2°C of global warming, a level beyond which scientists predict very disruptive or even 'dangerous' consequences (World Bank, 2010a).

In terms of future climate change, the IPCC Fourth Assessment Report compared the outputs of a wide range of different global climate models. It projects global warming of about 0.2°C per decade for 2007–2027 across a range of SRES emissions scenarios (similar to the observed recent decadal change). Beyond 2027, average temperature increases differ depending on which future emissions scenario is applied, ranging from a 1.8°C [1.1–2.9] rise for a B1 scenario to 4.0°C [2.4–6.4] for the A1FI scenario. Projected future warming is unevenly distributed, with the greatest changes over land and most high northern latitudes, and the least over the Southern Ocean (near Antarctica) and the northern North Atlantic.

Figure 1.4 *Changes in temperature, sea level and Northern Hemisphere snow cover. Observed changes in: (a) global average surface temperature; (b) global average sea level from tide gauge and satellite data; and (c) Northern Hemisphere snow cover for March–April.*

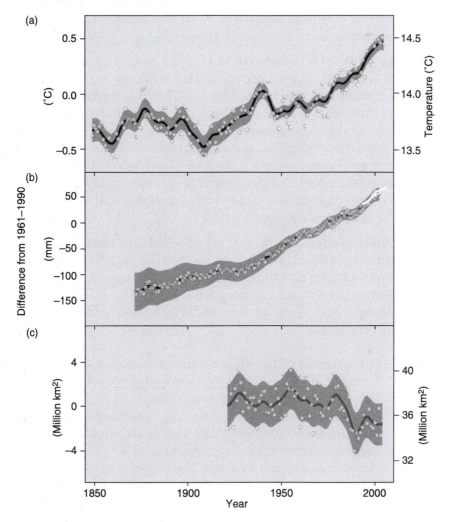

Source: IPCC, 2007a.
Note: All differences are relative to corresponding averages for the period 1961–1990. Smoothed curves represent decadal averaged values while circles show yearly values. The shaded areas are the uncertainty intervals estimated from a comprehensive analysis of known uncertainties [(a) and (b)] and from the time series (c).

This warming will shrink sea ice and snow cover and thaw permafrost. It is very likely that hot extremes, heat waves and heavy precipitation events will become more frequent, while future tropical cyclones (typhoons and hurricanes) are likely to become more intense, with larger peak wind speeds and more heavy precipitation. Increases in the amount of precipitation are very likely in high latitudes, while decreases are likely in most subtropical land regions. This decrease may be by as much as 20 per cent in the A1B scenario by 2100.

As a result of both thermal ocean expansion and melting land ice, the rate of global sea level rise is projected to increase during the twenty-first century, reaching an average of 0.22 to 0.44 metres above 1990 levels by the mid-2090s under the A1B emissions scenario, although this will vary geographically (IPCC, 2007a). This projection should be considered conservative, as it does not account for the possible melting of Greenland and Antarctica's ice-sheets. The projected rises are significant given the number of people, amount of infrastructure, productive agricultural land and other economic activity located in low-lying coastal areas.

Climate change will also alter the magnitude and frequency of some extreme weather and climate events. In 2011, the IPCC launched its 'Special Report on Managing the Risks of Extreme Events and Disasters to Advance Climate Change Adaptation' (SREX). Key messages from the report summarized by Mitchell and van Aalst (2011) include:

1 Even without taking climate change into account, disaster risk will continue to increase in many countries as more people and assets are exposed to weather extremes.
2 Evidence suggests that climate change has changed the magnitude and frequency of some extreme weather and climate events ('climate extremes') in some regions already.
3 Climate change will have significant impacts on the severity and magnitude of climate extremes in the future, especially as climate change becomes more dramatic.
4 High levels of vulnerability, combined with more severe and frequent weather and climate extremes, may result in some places, such as atolls, being increasingly difficult places in which to live and work.
5 A new balance needs to be struck between measures to reduce risk, transfer risk (e.g. through insurance) and effectively prepare for and manage disaster impact in a changing climate. This balance will require a stronger emphasis on anticipation and risk reduction.
6 In this context, existing risk management measures need to be

improved as many countries are poorly adapted to current extremes and risks, let alone those projected for the future.

7   In cases where vulnerability and exposure are high, capacity is low, and weather and climate extremes are changing, more fundamental 'transformational' adjustments may be required to avoid the worst disaster losses.

8   Any delay in greenhouse gas mitigation is likely to lead to more severe and frequent climate extremes.

Crucially, the possibility that climate changes may be abrupt rather than gradual is a major cause for concern. Andrew Dessler (2012: p. 149) likens this difference to the difference between gradually rocking a canoe and surpassing a critical threshold where it flips over, throwing you into the water. Examples include disruption to significant regional climate systems relevant to developing countries such as shifts to the monsoon in South Asia and West Africa, and to the El Niño Southern Oscillation effects on climate in the Pacific Rim and beyond. Historical records suggest that these changes have occurred in the geological past, particularly through disruption to ocean circulation, with major implications for regional climates.

6. Regional impacts of climate change in the developing world

Although the impacts of global climate change are highly location-specific, the IPCC determines a range of projected impacts of varying magnitude depending on future atmospheric concentrations of greenhouse gases and associated temperature rise (Figure 1.5). This section summarizes some of the projected regional impacts of climate change pertinent to developing countries based on the IPCC's Fourth Assessment Report (IPCC, 2007a). In general, climate research and information on developing countries and regions lags behind industrialized countries and regions, with many developing countries facing significant capacity challenges (see Box 1.5).

In Africa, between 75 and 250 million people are projected to be exposed to increased water stress due to climate change by 2020. Agricultural production is projected to be severely compromised in many African countries, adversely affecting food security and exacerbating malnutrition. By 2020, yields from rain-fed agriculture in

Figure 1.5 Projected impacts of climate change by region.

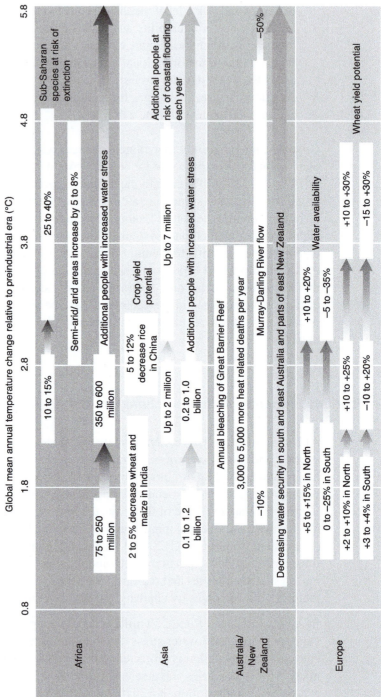

Global mean annual temperature change relative to preindustrial era (°C)

Latin America

Many tropical glaciers disappear

10 to 80 million

80 to 180 million

Many mid-latitude glaciers disappear

Potential extinction of about 25% Central Brazilian savanna tree species

Potential extinction of about 45% Amazonian tree species

North America

5 to 20% increase crop yield potential

Decreased space heating and increased space cooling

Additional people with increased water stress

About 70% increase in hazardous ozone days

70 to 120% increase forest area burned in Canada

3 to 8 times increase in heat wave days in some cities

Polar Regions

Increase in depth of seasonal thaw of Arctic permafrost

10 to 15%

15 to 25%

20 to 35% reduction of Arctic permafrost area

30 to 50%

10 to 50% Arctic tundra replaced by forest

15 to 25% polar desert replaced by tundra

20 to 35% decrease annual average Arctic sea ice area

Small Islands

Increasing coastal inundation and damage to infrastructure due to sea-level rise

Alien species colonize mid- and high latitude islands

Agricultural losses up to 5% GDP in high terrain islands, up to 20% GDP in low terrain islands

0   1   2   3   4   5

Global mean annual temperature change relative to 1980–99 (°C)

Source: World Bank, 2010a, adapted from Parry et al., 2007.

Box 1.5

Capacity and climate science in developing countries

Developing countries face a range of challenges relating to climate information and climate science. These include:

● Limited and sometimes unreliable historical datasets both on climate and on emissions

● Ongoing paucity of meteorological and emissions data collection

● Limited development of climate models, particularly regional climate models, and emissions methodologies

● Limited research efforts to develop and apply models for developing country contexts

The availability of data and analysis, and the capacity to advance scientific understanding, varies significantly around the world. Climate data records are generally longer and better in the more industrialized nations, while in some parts of the world, especially Sub-Saharan Africa, there are large gaps in historical data.

Climate change expertise is also unevenly distributed and unequally represented in global assessment processes such as the IPCC, even when differences in capacity, population and economic development are taken into account. In their study of the first four IPCC Assessment Reports, Ho-Lem *et al.* (2011) found that 45 per cent of countries, all of them developing countries, have never had authors participate in the IPCC process. In contrast, European and North American experts make up more than 75 per cent of IPCC authors.

some countries could be reduced by up to 50 per cent. Towards the end of the twenty-first century, projected sea level rise will affect low-lying coastal areas with large populations. The cost of adaptation could amount to at least 5 per cent to 10 per cent of Africa's Gross Domestic Product.

Climate change will compound the pressures on natural resources and the environment associated with rapid urbanization, industrialization and economic development in Asia. Coastal flooding is projected to increase, especially in heavily populated mega-delta regions in South, East and South-East Asia, where sea-effects combine with greater river flows. Freshwater availability in Central, South, East and South-East Asia, particularly in large river basins, is projected to decrease by the 2050s. Morbidity and mortality due to diarrhoeal disease primarily associated with floods and droughts are expected to rise in East, South and South-East Asia due to projected changes in the hydrological cycle.

Tropical Latin America faces risk of significant biodiversity loss through species extinction in many areas. Semiarid vegetation will tend to be replaced by arid-land vegetation, and the productivity of some important crops is projected to decrease and livestock productivity to decline, with adverse consequences for food security. At the same time, soybean yields in temperate zones are projected to increase. Changes in precipitation patterns and the disappearance of glaciers are projected to affect significantly water availability for human consumption, agriculture and energy generation.

In Small Islands, sea level rise is the chief concern and is expected to exacerbate inundation, storm surge, erosion and other coastal **hazards**, threatening vital infrastructure, settlements and facilities that support the livelihood of island communities. Deterioration in coastal conditions, for example through erosion of beaches and coral bleaching, is expected to affect local resources. By mid-century, climate change is expected to reduce water resources in many small islands to the point where they become insufficient to meet demand during low-rainfall periods.

7. From climate science to policy response

In the previous sections we summarize the significance of climate change for the future of the planet and human development. It follows that we need to determine what actions to take to limit and cope with its impacts. Deciding how to respond to climate change depends on a complex mix of scientific and non-scientific factors, scientific evidence and economic calculations, but also on values and political judgements, including domestic interests and international agreements on tackling climate change (see Chapter 3).

Stabilizing atmospheric greenhouse gas concentrations at any given level could be achieved by a variety of emissions pathways. For any given concentration level (measured in *parts per million of CO_2 equivalent*, often shortened to *ppm*), the earlier that emissions levels peak, the slower they then need to be reduced. In other words, the greater the delay in starting to reduce emissions, the faster the rate of reduction required and the greater the likely expense of doing so. These different 'windows of opportunity' are used to support arguments for early action to reduce emissions (Stern, 2007; UNDP, 2007). For example, for the target concentration level of 450 ppm, if emissions reductions begin now, the required emissions reductions rate is around 1.5 per cent per year. If we delay reductions by 8 to 10 years, however, this pushes the required reduction rate to greater than 3 per cent per year – a rate widely regarded as beyond current technological means (Mignone *et al.*, 2008; this topic is examined further in Chapter 4).

An important top-level decision is about the extent to which we should act to reduce greenhouse gas concentrations (mitigation) or whether we accept impacts and adapt to them (adaptation). Both entail a cost; spending more on mitigation implies spending less on adaptation and vice versa. Using cost-benefit analysis to determine the optimum mix of mitigation and adaptation, economists make a series of assumptions regarding a number of factors that include: Climate sensitivity (the degree of climate change that results from a given increase in greenhouse gases); mitigation costs (especially regarding the availability and costs of future technology); the value of damages from impacts (depending on both the increased burden of hazards and the changing vulnerability of the system); costs of adaptation (what reduction in damages can be achieved for a given adaptation action); and discount rates (the difference between how much we value costs and benefits now compared to in the future) (World Bank, 2010a; this topic is examined further in section 4.1 of Chapter 2).

These analyses also rely on normative value judgements about how to incorporate and measure the distribution of actions across time (mitigation may cost now but will benefit future generations), the distribution of impacts (impacts are often treated as a global total but in reality they will vary across the world's regions and people), and whether to value non-economic factors such as the loss of life, livelihoods, culture and nonmarket services such as species extinction, biodiversity and ecosystem services.

While our actions can draw on the evidence presented by science and economics, ultimately judgements are based on the politics and values framing different decision-making contexts (Toman, 2006; O'Brien, 2009; Dessler, 2012). For example, the dominant global climate change response under the UNFCCC sets the goal as avoiding 'dangerous interference' with the climate system (see Chapter 3).

However, what is considered dangerous is to a large extent a societal value judgement rather than a scientific one and we tend not to act on risks until they become real, even if those risks may be significant and irreversible. Different values held by different people also influence how they will act. For example, Dessler (2012: p. 216) contrasts two opposing views of how to respond to climate change, even though both use the precautionary principle as their basis:

- Because the worst-case scenario of climate change is so serious, we must take action now to reduce emissions, even though we do not know exactly how bad climate change will be.
- Because of the high cost of reducing emissions, we must be certain that climate change is serious before we take action.

Our lived experiences and the way we receive information about climate change also influence our responses to climate change. Political beliefs and values have been found to play a key role in determining whether we accept or deny the links between human actions and climate change (Whitmarsh, 2011). Decisions about how to respond to climate change may also be influenced more by the views of key opinion leaders or prevailing media representations than the objective evidence provided by scientific experts. Most of us do not directly access scientific experts and journals; we rely on the media and thought leaders to read, digest and communicate them for us. There is therefore a degree of subjectivity built into the very process by which we learn about climate science and impacts.

Conclusion

Our understanding of climate science is evolving rapidly. Despite the attempts of climate sceptics to dispute the science, there is overwhelming evidence that human activity is changing the climate system. Industrial development has driven human-caused climate change since the beginning of the industrial era. Climate change impacts will be felt in all parts of the world, but they will differ across regions. While research in and focus on developing countries has lagged behind other parts of the world, there is growing understanding that climate change impacts will be felt acutely there. Chapter 2 explores the nexus between the processes of development and our responses to climate change.

Summary

- Climate science has a privileged position in informing debates and action. Consequently, a rudimentary knowledge of atmospheric systems and climate science is essential for understanding the interactions between climate change and development.

- The counter-claims of climate sceptics are important in so far as they influence behavioral and policy responses to climate change, even though they generally rely on a scientifically weak or partial evidence base.

- The global climate system comprises a complex interaction of the oceans, land surface and atmosphere. Rising greenhouse gas concentrations in the atmosphere have enhanced the natural 'greenhouse' effect by absorbing outgoing infrared radiation in the atmosphere.

- Global surface temperatures today are 0.89°C above pre-industrial levels. Warming of the climate system is unequivocal and human activity is largely responsible. Human activity is increasing concentrations of greenhouse gases in the atmosphere and changing the earth's land surface. The observed warming cannot be attributed to known natural causes alone.

- The impacts of climate change will vary by region. Water stress, especially to rain-fed agriculture, inundation of low lying coastal areas and small islands, and health impacts will all have profound implications in developing countries, where research on climate science has lagged behind other regions.

- Decisions about how to respond to climate change are informed by the science of impacts and atmospheric stabilization, but also by politics and economics of the costs and benefits of action and inaction. Underlying values and beliefs will affect our attitude to these factors.

Discussion questions

- Why are the claims of climate sceptics and deniers so influential despite their lack of scientific credibility?

- How are human activities other than those that generate or reduce greenhouse gases affecting the climate system?

- What sorts of decision-making are most appropriate for climate change model outputs? What other types of climate information might be appropriate to different contexts in developing countries?

- What are the priorities for improving climate change research capacity in developing countries?

- What are the implications of politics and values for developing countries in deciding the balance between how much to mitigate climate change and how much to adapt to its impacts?

Further reading

Dessler, A.E. (2012) *Introduction to Modern Climate Change*, Cambridge: Cambridge University Press.

Hulme, M. (2009) *Why We Disagree About Climate Change: Understanding Controversy, Inaction and Opportunity*, Cambridge: Cambridge University Press.

IPCC (2007) *Climate Change 2007: Synthesis Report. An Assessment of the Intergovernmental Panel on Climate Change*, Cambridge, UK, and New York, NY, USA: Cambridge University Press.

Schneider, S. (2009) *Science as a Contact Sport: Inside the Battle to Save Earth's Climate*, Washington DC: National Geographic Press.

World Bank (2010) *World Development Report: Development and Climate Change*, Washington, D.C.: The World Bank [see Focus A – 'The Science of Climate Change'].

Websites

www.ipcc.ch
Website of the Intergovernmental Panel on Climate Change (IPCC): the leading international body for the assessment of climate change. The site summarises their activities and links to IPCC Assessment Reports, methodological guidance, Technical Papers and Special Reports.

www.climate.gov
USA National Oceanic and Atmospheric Administration website. Links to many
videos and images, *ClimateWatch Magazine*, and updates on global climate change
science in the State of the Climate reports.

www.realclimate.org
A fairly technical website written by climate scientists to tackle scientific topics and
issues. It covers stories about new scientific papers, major events in the climate
debate, and hosts blog posts on aspects of the climate system.

www.wmo.ch
World Meteorological Organization, which coordinates global scientific activity in
weather and climate-related issues.

http://cait.wri.org/
Climate Analysis Indicators Tool (CAIT) provides a comprehensive and comparable
database of greenhouse gas emissions data and other climate-relevant indicators.

2 The climate–development nexus

Climate and development goals are deeply inter-twined.

Introduction

The previous chapter shed light on the nature of **climate change** primarily from a scientific perspective. In this chapter climate change is viewed through a societal lens, as a development challenge. We consider the complex interactions and deep interdependencies between our **climate** and socio-economic systems.

Increasingly, development efforts are at risk of being derailed by the vagaries of a changing climate. Continued socio-economic progress will depend on our ability to manage both climate risks and our influence on the climate. At the same time, development choices directly influence the climate, and the need to stabilize atmospheric greenhouse gas concentrations at a safe level will create profound challenges as well as unprecedented opportunities for development efforts in the future. Like an infinity knot, our climate and development futures are entwined in a circular chain of cause and effect (see Figure 2.1). This poses a profound challenge for governments, development organizations and development practitioners, whose thinking and practice has for long been conditioned by the assumption of linear causality.

The interdependency of our climate protection and development goals is underpinned by a set of complex interactions between climate, environmental and socio-economic systems, which we refer to as the

Figure 2.1 *The 'Infinity Knot' of Climate and Development.*

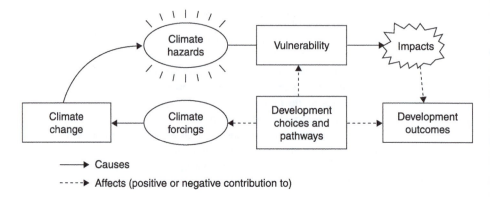

Source: the authors.

climate–development nexus. These interactions play out on multiple spatial and time scales, calling for international collective action and raising concerns of intra- and inter-generational **equity** and justice. Thus climate change and development are linked not just causally, but also normatively and conceptually: how we think about and act upon both is shaped by common principles and moral considerations, such as equity, **social justice** and sustainability.

This chapter explains why and how climate change matters to development, and development to climate change. It conceptually untangles the separate threads that together form the infinity knot of climate and development. Readers should gain an appreciation of why meaningful and sustained progress in addressing one requires concomitant progress in addressing the other, and therefore why climate change and development must be addressed together.

The first part of this chapter considers how ideas of equity, social justice and sustainability provide common normative framings for development and climate action. The second part examines how development choices affect our climate and how the need to stabilize the climate will have far-reaching implications for development choices and pathways. The third part examines how climate change affects development outcomes, and suggests a broad conceptualization of **vulnerability** as a key to understanding climate risks. The fourth part considers how climate and development responses inter-relate and how they can be bridged. This last section sets the scene for Section

Two of the book, which explores how climate change responses can be made more sensitive to development concerns, and Section Three, which focuses on how development efforts can be made compatible with climate protection.

1. Common normative framings

Climate and development challenges share several characteristics. They both require urgent action while being long-term projects spanning several generations. Both call for responses ranging from the local to the global level, and demand actions by multiple stakeholders, including governments, the private sector, and civil society. Actions in both spheres also have common normative underpinnings. Notions of equity, social justice and sustainability provide common moral justifications and the core principles that guide both climate protection and development. In this section, we examine how ideas of social justice, equity, and sustainability are interwoven into the climate and development threads of our infinity knot.

1.1 Social justice and equity

Social justice is best understood as a process, rather than an end point or outcome. More specifically, it is a process that (a) seeks a fair distribution of resources, opportunities and responsibilities; (b) challenges the roots of injustice; (c) empowers all people to fulfill their potential and lead lives they value; and (d) builds social solidarity and capacity for collective action (adapted from UC Berkeley 7th Annual Justice Symposium).

Equity is an ethical principle based on the notion of moral equality between individuals, i.e. that all should be treated as equals. Equity has three main dimensions: (a) equality of opportunity – a person's life chances should not be determined by the accident of birth; (b) equal concern for people's needs – ensuring that all are guaranteed basic rights to access goods and services fundamental to their needs and to preserving their human dignity; and (c) meritocracy – that is positions in society and rewards reflect differences in effort and ability and are based on fair competition (Jones, 2009).

Equity, social justice and development

From the outset, international development was driven by an understanding that international peace and stability depend on shared prosperity and an international order based on universal human rights and international distributive justice. The moral outrage at avoidable human poverty in a world of material abundance has also been an important motivator.

Since the 1980s, equity considerations had taken a back seat in international development discourse. This may be explained in large part by two factors: ideology and economic theory. First, in the charged and polarised ideological environment of the cold war, leaders of the 'free world' took a stand against 'the politics of envy' in their vow to defend freedom against totalitarianism (Fuentes, 2012). Second, the development discourse has been dominated by mainstream economists who have, until recently, largely held that equality obstructs efficiency. In his influential 1975 book *Equality and Efficiency: The Big Tradeoff*, economist Arthur Okun argued that pursuing equality can reduce efficiency in the economy by reducing incentives to work and invest, and also because efforts to redistribute – through such mechanisms as the tax code and minimum wages – can themselves be costly. Robert Lucas, a Nobel Prize winning economist, opined that 'nothing [is] as poisonous to sound economics as to "focus on questions of distribution" '.

More recently, a flood of evidence has shown that too much inequality impedes socio-economic and political progress, and correcting inequity can deliver sizeable and long-lasting social and economic returns (Acemoglu, 2012; Berg and Ostry, 2011; Fuentes, 2012; Stiglitz, 2012; Wilkinson and Pickett, 2010). Health researchers Richard Wilkinson and Kate Pickett argue forcefully in *The Spirit Level* that almost every aspect of well-being – life expectancy, mental health, violence, illiteracy, social and political inclusion – is affected less by how wealthy a society is than by how unequal it is.

Contrary to Okun's theory, there is evidence that some inequalities – in access to education, credit, and land – hamper growth because the talents, energy and ideas of a large part of the population are underused, creating simmering frustration and social discontent (Evans, 2012). Recent work by the International Monetary Fund reveals that when growth is viewed over the long term, trade-offs

between efficiency and equality may be nonexistent. In
fact, greater equality appears to promote and sustain growth (Berg and
Ostry, 2011).

Equity, social justice and climate change

Equity and justice concerns have been central to the international
negotiations of the **UNFCCC** from the outset (see Chapter 3). Indeed,
any serious action on climate change confronts serious ethical issues
of fairness and responsibility between people, nations, and
generations. There are two main injustices at stake. The first is that, as
a result of delayed action by rich countries to curb their emissions,
developing countries have been forced into a situation where they
must commit to significant cuts in their own emissions, even though
they only contributed one-fifth of historical emissions between 1850
and 2002 (WRI, 2005). This is what Lord Stern called the 'brutal
arithmetic' of climate change. This is explained by the changing
structure of the world economy and the growing contribution of big
emerging economies (including China, India, Brazil) to global
emissions. Whereas developing countries contributed about a third
of global emissions in 1990 – the baseline year for the Kyoto
Protocol – they are expected to account for two-thirds of emissions
by 2030.

The second injustice is that the impacts of climate change fall most
heavily on those who contributed the least to global emissions: poor
people. It is estimated that developing countries will bear
approximately four-fifths of the costs of the damages caused by a
warming of 2°C (World Bank, 2010). Poor peoples' natural trust funds
– the fields, fisheries, forests and waterways on which they depend
directly – will be increasingly at risk because of climate change. The
inequities in the global distribution of emissions and impacts are
illustrated in Box 2.1.

Since meeting our internationally agreed climate **stabilization** goals
requires all countries, rich and poor, to contribute to **mitigation**
efforts, there is a very practical reason for ensuring equity in climate
change responses: measures to cut emissions need to be fair if they are
to gain widespread acceptance. But as we know, emissions are
unequally distributed: the per capita energy-related carbon footprint is
nearly 12 times higher in rich countries, and more than three times
higher in middle-income countries, than in poor countries (World

Box 2.1

Climate injustice: unequal causes and impacts of climate change

Here's what the world looks like if country sizes were proportional to their emissions (world map scaled to fossil-fuel carbon-dioxide emissions in 2010):

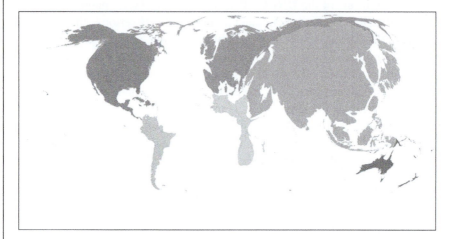

Taking account of historical emissions, measured as cumulative CO_2 emissions from energy use between 1850 and 2008, gives us the following map:

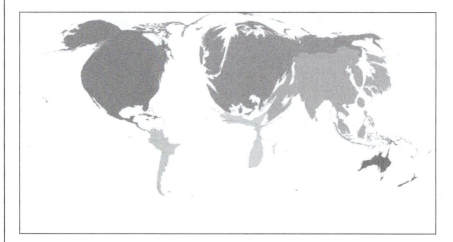

And here's what the world looks like if countries were sized commensurately with the number of people at risk from the impacts of climate change (world map scaled by number of people exposed to droughts, floods and extreme temperatures in 2010):

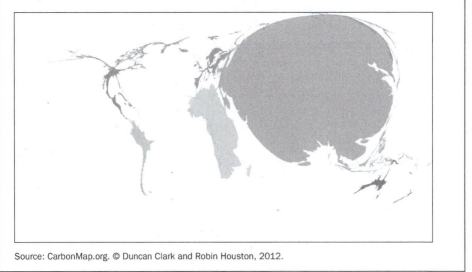

Source: CarbonMap.org. © Duncan Clark and Robin Houston, 2012.

Bank, 2010). The available **carbon space** – that is the total amount of CO_2 emissions that can be allowed if stabilization targets were taken seriously – should be shared in a way that is fair, which requires 'convergence' of currently widely unequal per capita emissions. As noted in the World Development Report 2010, 'it is ethically and politically unacceptable to deny the world's poor the opportunity to ascend the income ladder simply because the rich reached the top first' (World Bank, 2010; p. 44).

At low levels of **human development**, increased **greenhouse gas** emissions can be considered a human right as they are linked to the fulfillment of basic human needs (e.g. energy access, development of basic infrastructure etc). In contrast, increased emissions in rich countries are often associated with luxurious consumption (Pan, 2004). Fairness dictates a differentiated approach to addressing greenhouse gas emission in these different situations.

Issues of equity also arise in addressing the impacts of climate change, particularly considering countries' differing abilities to cope with and adapt to a changing climate. Three kinds of inequality exacerbate poor people's vulnerabilities to climate risks. First, economic and legal inequalities affect poor people's access to the resources needed to cope with shocks (anticipated and unanticipated). Second, infrastructural disparities mean that the poorest are most sensitive to and least equipped to withstand extreme weather events. Third, the unequal distribution of power and influence over decision-making means that the allocation of public resources tends to favour the rich, even in times of crisis. A quantitative study by Roberts and Parks (2007) confirms that 'countries with high levels of income inequality experience the effects of climate disasters more profoundly than more equal societies'.

1.2 Sustainability

Definitions

Like equity and social justice, the notion of sustainability is tightly intertwined in how we think about and act on climate change and development. In a strict sense sustainability can be understood as the property of any system to maintain its performance over time (Fuentes and Pereira, 2010). Here, the performance of interest is development. As noted in the introduction, we broadly define development as the improvement of well-being through the enlargement of human freedoms and capabilities.

Well-being is difficult to measure directly at an aggregate level. To circumvent this challenge, economists rely on proxies, such as income, consumption and capital, defined as a stock of wealth from which economically desirable outputs are generated. In this approach, sustainability is framed as the maintenance of a certain level of capital wealth over time.

Five kinds of capital are essential for sustainable development:

- *Financial capital* facilitates economic production and exchange, though it is not itself productive. It refers to financial assets, such as cash, equity and debt.
- *Human capital* refers to the productive capacities of an individual – knowledge, skills and abilities – both inherited and acquired through education, training and socialization.
- *Social capital* consists of networks and interactions that breed social cohesion, trust, and creativity.
- *Natural capital* is made up of the resources and ecosystem services of the natural world. It includes resources like minerals, oil, or

wood; and regulating services provided by intact ecosystems, such as storm surge protection from mangrove forests.

- *Physical capital* consists of physical assets generated by applying human productive capacities (human capital) to transforming natural capital into infrastructure, equipment, and technologies (e.g. roads, power plants, and buildings).

There are two schools of thought on sustainability as maintaining the different types of capital (Fuentes and Pereira, 2010):

(i) The weak sustainability school, which holds that the different types of capital are substitutes for each other and the main concern should be the total stock of capital (produced, human, social and natural). In this school, sustainability translates to a non-decreasing real value of the total stock of capital.

(ii) The strong sustainability school, which holds that some types of capital are critical, namely natural capital (or parts of it), and cannot be substituted by other types of capital. In this school, sustainability means that the stock of the critical capital(s) does not decrease.

Sustainability and climate change

Climate change is an inter-generational challenge. It requires costly actions today for the sake of generations to come, and precludes poor countries today from following the path to prosperity that rich countries trod in the past.

Concern about climate change grew out of a broader concern regarding the unsustainability of our development trajectories. Industrialization has made possible improvements in well-being on an unprecedented scale. In light of climate change, however, the development path rich countries have followed – characterized by heavy reliance on carbon-rich energy sources and resource-intensive modes of production and consumption – cannot be sustained.

Over the last 50 years, we have changed ecosystems more rapidly than at any time in human history, largely to meet growing demands for food, freshwater, timber, and fiber. Two-thirds of the ecosystem services humans depend on are degraded (Millennium Ecosystem Assessment, 2005). According to WWF, the rate at which we are currently using up the Earth's resources exceeds the planet's capacity to regenerate them by 50 per cent (WWF, 2012). As Figure 2.2 shows, even though there have

Figure 2.2 *Rising resource consumption despite improvements in resource use efficiency.*

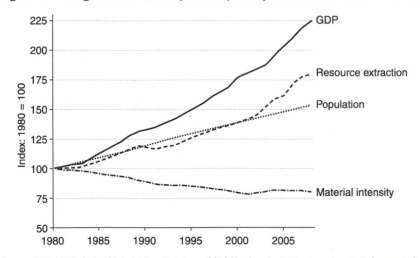

Source: SERI 2012. Global Material Flow Database. 2012 Version. Available at: www.materialflows.net. Sustainable Europe Research Institute, Vienna, Austria.

been impressive improvements in the efficiency of resource use, the total consumption of resources has been rising steadily.

Greenhouse gas emissions and environmental trends mirror each other. They share common drivers: both are symptoms of environmentally unsustainable economic expansion and a mode of development that externalizes pollution and environmental costs (Daly, 1999). They are also causally related and mutually reinforcing: climate change is set to become a major driver of biodiversity loss and ecosystem degradation (Millennium Ecosystem Assessment, 2005) and conversely, the loss of carbon-rich ecosystems such as rainforests and peat-lands has the potential to contribute significantly to global warming.

Sustainability and development

Our collective actions today will determine the kind of societies and environment that future generations inherit. Failure to curb global warming will limit the opportunities and freedoms available to generations to come.

Environmental stewardship matters not only for sustaining the material basis to meet the needs and aspirations of future generations, but also because of the immediate impact of environmental factors on peoples' well-being. Environmental conditions affect people both directly (through exposure to pollution, environmental **hazards** and vector-borne diseases) and indirectly (through climate risks and the degradation of ecosystems on which people depend for food, fiber, fuel and income).

For example, 35 per cent of total household income in Zimbabwe can be considered environmental income – tied to crops, livestocks, woodlands and gardens (WRI, 2005). A corollary of dependence is that poor people are the most acutely vulnerable to the depletion and degradation of environmental resources. Indeed the activities most affected by the loss or degradation of ecosystems are subsistence farming, animal husbandry, fishing and informal forestry, all of which are key livelihoods for poor people. Also, environmental factors – such as air and water pollution – are among the top causes of sickness and death, contributing 20 per cent of the disease burden in developing countries.

From a 'capitals' perspective, reducing poverty involves protecting and growing peoples capital stock. This requires tackling the inequities deeply embedded in the distribution of the various forms of capital, which relates to the way institutions are governed and decisions made. Beyond asset accumulation, breaking through the **poverty trap** requires multi-pronged approaches that tackle the socio-economic, environmental and institutional root causes of poverty. The Sustainable Livelihoods Approach is a widely used framework for analyzing how the five forms of capital are impacted by external shocks and stresses, and how institutions and policies frame our responses (see Box 2.2).

Box 2.2

The Sustainable Livelihoods Approach

The Sustainable Livelihoods Approach (SLA) provides a useful framework for understanding how individuals' capital stock (or 'livelihood assets') is affected by external shocks and stresses, and how their capacity to respond to these is conditioned by policies, institutions and organizational processes. Developed in the 1990s from research led by the Institute for Development Studies, the SLA offers a framework for understanding the wider range of factors that cause people to fall into poverty. Whereas conventional understandings of poverty had focused on the factors that cause poverty in the aggregate across an entire economy, the SLA offered a people-centered approach, meaning that it focuses on the forces that drive individuals and communities into poverty and the assets they need to build resilience.

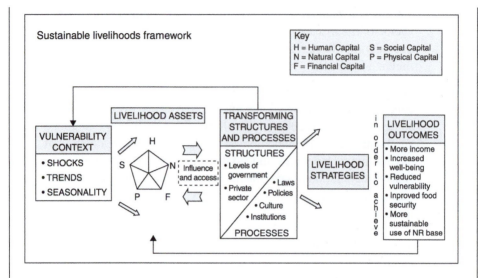

Illustrated above, the framework draws on five core concepts:

● **Vulnerability Context** describes the external conditions in which people live and breaks these conditions down into the following: 'shocks' – one-time events like extreme weather or an outbreak of an infectious disease; 'trends' – long-term changes in population or economics, for instance; and 'seasonality' – cyclical variations in prices, employment, production, or health.

● **Livelihood Assets** define the strengths from which people seek to increase their well-being. The assets are categorized into five forms of capital: financial, human, social, natural and physical.

● **Transforming Structures and Processes** are the 'the institutions, organisations, policies and legislation that shape livelihoods' (DFID 1999: p. 17). These range from formal political institutions to corporations and the private sector to cultural conventions that shape social behavior in a community. The structures and processes help to define an individual or community's ability to access livelihood assets and therefore to employ livelihood strategies that achieve positive livelihood outcomes.

● **Livelihood Strategies** refer to the ways people use their livelihood assets to build well-being, while **Livelihood Outcomes** are the results of these strategies.

Source: Conway and Chambers, 1991; DFID, 1999.

2. Carbon space, mitigation and development

2.1 Climate change mitigation and carbon space

Climate change mitigation is the attempt to reduce the rate at which greenhouse gases accumulate in the atmosphere, in order to minimize climate change and its effects. This is done by reducing greenhouse gas emissions or removing greenhouse gases from the atmosphere through enhancing carbon sinks (e.g. trees, soil, vegetation). Mitigation includes all actions aimed at slowing the rate of climate change by reducing the emissions of **GHGs** such as carbon dioxide (CO_2) and methane (CH_4). Development that results in lower emissions of GHGs compared with the baseline, business-as-usual development path is often referred to as **low-carbon development**, or **low-emissions development**.

The objective of international climate change mitigation efforts, as set out in the articles of the UNFCCC, is to avert 'dangerous climate change'. What counts as 'dangerous' is of course subjective and relative given the asymmetric impacts of climate change. Drawing on the recommendations of the Intergovernmental Panel on Climate Change (IPCC), several countries and advocates have urged a global goal of restricting global temperature rise to 2°C (3.6°F) above pre-industrial levels, which will require reducing global emissions by at least 50 per cent by 2050 from their 2000 levels. Even this level of ambition is too low from the perspective of the most vulnerable nations. Thus the Alliance of Small Island States (AOSIS), whose very survival is at stake because of rising sea levels, have advocated for a more stringent target of 1.5°C.

The objective of stabilizing temperature increases at a given level establishes an upper limit on total allowable emissions from all sources over time, a quantity known variably as carbon space or **carbon budget**. Even a 2°C target would entail unprecedented reductions in greenhouse gas emissions, at a time of increasing affluence and a growing global population. It will require stabilization at 450ppm which translates to a carbon budget of 1,456 Gt CO_2 (see Figure 2.3). That means that an annual average of 14.5 Gt CO_2 of emissions can be released into the atmosphere in the course of the twenty-first century. Annual emissions in 2013 were more than twice this level. A 1.5°C target would entail a global carbon budget of 1,100 Gt CO_2, which would require emissions reductions far beyond those currently contemplated by politicians, as well as large **negative emissions** (Friends of the Earth, 2011).

The timing of mitigation actions is important, as noted in Chapter 1 (section 7). Because greenhouse gases persist in the atmosphere and because it is the stock – i.e. concentrations – of greenhouse gases in the atmosphere that matters, the more mitigation efforts are delayed,

Figure 2.3 *The 21st century carbon budget.*

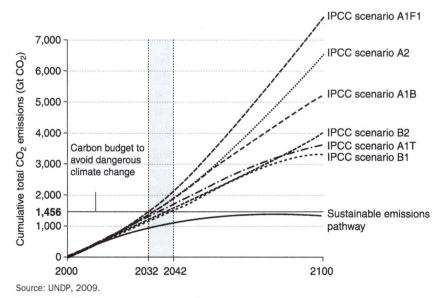

Source: UNDP, 2009.

Note: IPCC scenarios (SRES) describe plausible future patterns of population growth, economic growth, technological change and associated CO_2 emissions (see Box 1.4 in Chapter 1). In order to have a fifty-fifty chance of keeping global temperature rise below 2°C, no more than 1,456 Gt of carbon dioxide emissions should be released into the atmosphere in the 21st century. Our maximum carbon budget for the 21st century could therefore expire as early as 2032 if the world remains on the worst-case IPCC emissions scenario course (Curve 1) and by 2042 at the best-case scenario (Curve 6).

the greater the rate of emissions reductions needed to achieve the same concentration levels. If actions were taken today, the target concentration level of 450 ppm could be realized through emissions reductions of ~1.5 per cent per year. However, delaying actions by 10 years would mean that future annual emissions reductions of more than 3 per cent would be required to realize the target concentration level, a reduction rate widely regarded as beyond current technological means (Mignone *et al.*, 2008). As Figure 2.3 shows, our maximum carbon budget for the twenty-first century could expire as early as 2032 unless urgent action is taken to reduce emissions.

Because the emission of greenhouse gases is deeply rooted in how and what we produce and consume, there is a strong moral case that all humans are equally entitled to the global atmospheric commons. This view of the carbon space as a vital shared resource re-casts the issue of climate change mitigation as a matter of resource sharing, rather than one of cost-minimisation or burden-sharing.

Box 2.3

How to determine fair allocations of carbon space between countries

The notion of converging towards equal per capita entitlements is a core principle for equitable sharing of carbon space. A number of other considerations also need to be considered. There are four possible approaches to adjusting the per capita allocations, depending on different definitions of fairness:

- account for historical emissions (i.e., the United States and European Union have already used far more than an equal share of the global carbon budget, which starts in 1850, not 2010);

- account for development needs (i.e., some countries have exceptionally pressing needs for more energy use);

- account for geography (some countries need a greater allocation as they need more energy for heating or cooling);

- account for lock-in (some countries are locked in to high carbon economies, and need time to unlock themselves).

Source: Friends of the Earth, 2011.

There are different ways to determine what constitutes a fair way of sharing the available carbon space (see Box 2.3). Fairness dictates that differentiated mitigation pathways will be taken to keep within the global carbon budget. The most credible proposals call for deep and immediate reductions in rich countries in order to create carbon space for low-emitting developing countries to continue growing from their very low emissions baseline, while high-emitting emerging economies such as China, India and Brazil go for an early peak followed by a gradual decline in their emissions.

2.2 Why mitigation matters to developing countries

Mitigation will require large-scale investments and entail upfront costs that can be significant. But for most developing countries, the payoff for their individual efforts is trivial, distant and uncertain. This raises the question: what's in it for developing countries?

There are three main reasons why mitigation matters to developing countries. First, developing countries have the greatest stake in effective collective action to curb global warming because they bear the brunt of climate impacts. Second, mitigation actions at the national level may in some cases deliver beneficial social and environmental side effects (known as 'co-benefits') on the cheap because such actions would effectively be subsidised through international transfers. Third, by embracing low-carbon economies, developing countries would be better prepared to prosper in a carbon-rationed world.

Achieving agreed mitigation objectives will require a truly collective international effort. Even if rich countries were to bring their emissions down to zero the world would still be off course to meet the 2°C stabilisation target (see Figure 2.4). By 2050, eight billion of the projected nine billion people on the planet will be living in today's developing countries; their share of global emissions will be 70 per cent (World Bank, 2010: 203). Effective mitigation therefore depends on the participation of developing countries.

While it is true that most poor countries have contributed very little to climate change, their role addressing it can be significant. For example, Africa is one of the world's most important reservoirs of soil and other terrestrial carbon, estimated to account for at least 20 per cent of the

Figure 2.4 *Rich countries acting alone will not do the job of averting climate disaster.*

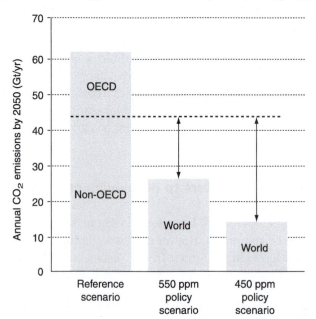

Source: World Bank, 2010a: 215.

world's entire stock of forest carbon and at least as great a share of its agricultural carbon, with very large potential for additional sequestration and other mitigation efforts. The world will not be able to limit global temperature increase to below 2°C without reducing emissions from land use and fully leveraging the remaining untapped capacity of ecosystems to store carbon; home to most of the world's rainforests, DR Congo, Brazil and Indonesia have particularly important roles to play in that process.

There can be significant social and environmental co-benefits from some mitigation actions. Alleviating pressures on scarce resources and reducing environmental pollution tends to benefit the poorest. For example, there is considerable evidence showing that most efforts to shift to clean cooking also reduce greenhouse gas emissions. Thus climate change mitigation can provide the impetus for investments in cleaner energy sources and technologies that deliver strong development co-benefits, such as improved local air quality and reduced household energy costs (IPCC Fourth Assessment Report, 7.10). Aside from the benefits derived from minimizing climate risks, developing countries can benefit from technology, financing and technical assistance available for participating in global efforts to curb emissions.

There are clear equity and efficiency reasons for developed countries to support developing countries' mitigation efforts. As already noted, since developing countries contributed the least to the problem and have the fewest resources to invest in mitigation, equity dictates that developed countries help finance mitigation in developing countries. At a global level, there is also a strong efficiency case for external financing of mitigation efforts in developing countries. Given the economic structures of developing countries, the low carbon efficiency of outdated technologies, and weak environmental controls, the marginal costs of abatement tend to be much lower in developing countries. In 2009, McKinsey & Company published a study surveying the emissions reductions potential from various economic sectors and regions throughout the world. The study found that 70 per cent of the most economically feasible emissions reductions opportunities are in developing countries (McKinsey & Company, 2009).

The benefits of national mitigation actions in developing countries are likely to be greatest in the longer term. First, by delinking economic growth and emissions increases, developing countries can expand their room to develop in what is likely to be an increasingly carbon-constrained world. Second, by reducing their dependence on fossil fuels, developing countries can be better insulated from international oil and gas price fluctuations.

Evidence from the oil price spikes of 2007–08 shows that the impact on poor countries' balance of payments and on poverty can be very severe (IMF, 2008). Because the current global food system is highly fuel- and transport-dependent, fuel increases also have a direct knock-on effect on international food prices, which in turn can have devastating consequences for poor people (see also Box 7.1). Developing countries that are highly dependent on imported fossil fuels will be penalized in a world of more expensive fossil energy, whether this results from growing scarcity ('peak oil') or from the introduction of an international carbon price.

3. Climate impacts, vulnerability and adaptation

Another way to reduce the impacts of climate change is to improve people's ability to cope with these impacts when they occur. This kind of response strategy is known as climate change **adaptation**. Climate change adaptation entails adjustments in natural or human systems in response to actual or anticipated changes in climatic conditions and their effects (IPCC, 2001). More recently, adaptation has been related to the concept of **resilience**, which helps communicate the idea of adaptation as a process of improving the ability to bounce back from shocks, among other characteristics (Bahadur *et al.*, 2013).

Chapter 1 explained that climate change manifests in the increasing incidence and intensity of climate-related weather extremes, and gradual changes in the mean conditions. In this section we explore the complex causal chains through which these climate hazards influence development.

3.1 Impacts on development

According to the IPCC, the consequences of increased temperatures are already being observed in natural systems around the world. According to the IPCC major sectors/resources affected by climate change include:

- **Water**: Water resources, which are critical to livelihoods, agriculture and power supply, are directly affected by climate.

Even as heavy precipitation events and flooding increase in frequency, drought-affected regions will likely expand. Water supplies stored in glaciers and snow cover, which serve more than one-sixth of the world population, will decline. Water is the major vulnerability in the Middle East and North Africa, the world's driest region, where per capita water availability is predicted to halve by 2050 even without the effects of climate change.

- **Ecosystems**: 20 to 30 per cent of plant and animal species are likely to be at risk of extinction if global average temperature increases exceed 1.5–2.5 degrees Celsius. Changes in ecosystem structure and function will have negative consequences for ecosystem goods and services such as water and food supply.
- **Food production**: 40 per cent of the world's population – and an even larger share of the world's poorest – relies on dry land agriculture in arid and semi-arid regions that are particularly vulnerable to drought as a result of climate change (FAO, 2012: 12). Regional shifts in the distribution and production of certain fish species will adversely affect aquaculture and fisheries.
- **Health**: Climate change poses a significant risk to human health, not only from increasing frequency of extreme weather events, but also from the increasing prevalence of infectious disease, especially in tropical regions (WHO, 2011: 25). Heat waves, floods, storms, fires and droughts are likely to claim more lives. The burden of cardio-respiratory and diarrheal diseases will increase and the spatial distribution of some infectious disease vectors will shift.
- **Infrastructure and human settlements**: Climate change can have devastating impacts on infrastructure and urban and rural settlements through sea level rise, shortage of water resources, extreme events, food security, health risks and temperature related morbidity. The bigger threat of climate variability to infrastructure is expected from rapid-onset disasters like storm surges, flash floods and tropical cyclones. Risks will be highest in low lying and densely populated human settlements. In East Asia and the Pacific, a major driver of vulnerability is the large number of people living along the coast and on low-lying islands: over 130 million in China and roughly 40 million – or more than half the entire population – in Vietnam (World Bank, 2009).

Poor people and countries are often the most vulnerable to the impacts of climate change. Most poor countries have predominantly natural resource-based economies that are sensitive to climatic conditions, with little diversification of economic activity, export sectors dominated by a few commodities, and weak manufacturing and services sectors. Moreover, climate change tends to affect the economic sectors on which poor people and countries depend the most. For example, agriculture is the main pillar of economic activity for most poor countries, and climate change is therefore likely to increase food insecurity in these countries (see Figure 2.5). If current emissions paths continue, over time climate change may have broader impact on rural livelihoods, including increasing water scarcity and large-scale species extinctions that degrade ecosystem services on which many poor rural people rely directly.

While science is not yet able to conclusively link any given climate hazard to climate change, let alone anthropogenic climate change, evidence compiled by the IPCC suggests that climate change has already increased the magnitude and frequency of some extreme weather and climate events in some regions (IPCC 2011). And

Figure 2.5 *Projected changes in Agricultural Output potential due to climate change, 2000–2080.*

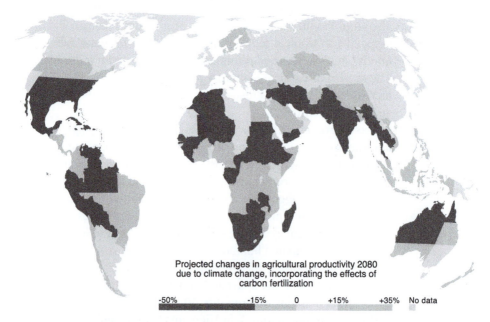

Projected changes in agricultural productivity 2080 due to climate change, incorporating the effects of carbon fertilization

-50%   -15%   0   +15%   +35%   No data

Source: Cline 2007, © Peterson Institute for International Economics, reproduced with permission.
Note: Of the 2.6 billion people who live on less than $2 per day, almost 2 billion live in rural areas, in countries whose economies and people are most dependent on natural resources. Efforts to meet the Millennium Development Goal of cutting poverty in half in such areas are being stymied by the already-evident impacts of climate change. Shown in this figure is the forecasted change in agricultural output potential from the turn of the century to 2080.

projected paths of climate change indicate a continued – and possibly accelerating – worsening trend. Recent science indicates that there is a very slim likelihood that the goal of stabilizing at 2 degrees will be met, and a 4 degree rise in temperatures is becoming more likely (New *et al.*, 2011; Stafford Smith *et al.*, 2011). At 4 degrees there is a higher risk of crossing **tipping points** in climate and large ecosystems. Crossing these tipping points will likely trigger very damaging climate-related impacts for which there is little or no precedent in human history. The consequences can be catastrophic, especially for poor people, and lead to large-scale development reversals.

3.2 The imperative for adaptation

Adaptation will be necessary regardless of how effective global mitigation efforts are. Given the inertia of the global climate system, a degree of warming is already unavoidable as a result of emissions to date: even if the global economy were to 'switch off' entirely for the next decade, current GHG concentrations already commit the world to increasing climate risks and impacts (see Chapter 1).

Adaptation is particularly important for poorer countries and poor people within developing countries because they generally:

- rely heavily on climate-sensitive sectors, such as agriculture and fisheries;
- are less able to respond to the direct and indirect effects of climate change due to limited **adaptive capacity**;
- are more exposed to climatic hazards – poor people are more likely to be located in marginal areas such as flood plains, steep hill slopes, or on nutrient-poor soils.

The imperative for adaptation in a developing country context is therefore partly driven by the practical need to reduce negative impacts on the processes of development, poverty reduction and growth (AfDB *et al.*, 2003; UNDP, 2007). As noted in the 2010 World Development Report 'even modest additional warming will require big adjustments to the way development policy is designed and implemented, to the way people live and make a living, and to the dangers and the opportunities they face' (World Bank, 2009).

The adaptation imperative is also fuelled by social justice concerns. As already discussed, developing countries have contributed the least but stand to lose the most from climate change. The fundamental unfairness in this reality has spurred NGO campaigns and demands from developing countries in the UNFCCC negotiations. In particular, the small island nations at risk from sea level rise and the least developed countries with low capacity to meet the costs of adaptation have stressed the need for compensatory measures to tackle climate change impacts (Burton *et al.*, 2002).

In many cases, although it may entail higher costs, climate change adaptation is consistent with and contributes towards development. Actions aimed at enhancing adaptive capacity and reducing the vulnerability of affected populations may also be good for development. Many adaptation measures simply entail rolling out proven development interventions, e.g. improving water storage, extending social protection measures to provide a buffer for poor people against climate-related shocks, distributing mosquito-proof bed nets to new areas that become exposed to the spread of malaria. Effectively these are 'low regret' actions, as their implementation is expected to benefit the poor irrespective of whether climate change occurs as projected.

Conversely, many non-climate-focused development interventions also contribute to enhancing capacities and reducing vulnerability to climate change. As noted by McGray *et al.* in their seminal report *Weathering the Storm*, 'rarely does adaptation entail activities not found in the development toolbox' (McGray *et al.*, 2007: 1). Developing countries will have good reason to prioritize **low-regrets measures** in tackling climate change, recognizing the significant development co-benefits of climate change adaptation. The range of approaches to adaptation, from pure development interventions that increase adaptive capacity (vulnerability focus) to interventions that address climate hazards directly (impacts focus) are summarized in Box 2.4.

Box 2.4

The development–adaptation continuum

In their 2007 report, *Weathering the Storm*, McGray *et al.* define an adaptation–development continuum to help understand the synergies and trade-offs between development strategies and climate change adaptation efforts.

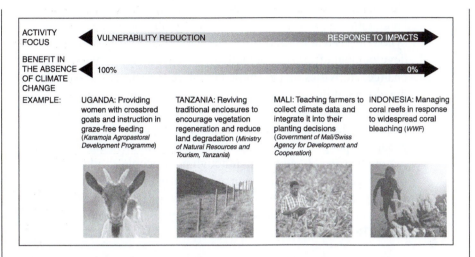

ACTIVITY FOCUS	◄ VULNERABILITY REDUCTION	RESPONSE TO IMPACTS ►		
BENEFIT IN THE ABSENCE OF CLIMATE CHANGE	◄ 100%			0% ►
EXAMPLE:	UGANDA: Providing women with crossbred goats and instruction in graze-free feeding (*Karamoja Agropastoral Development Programme*)	TANZANIA: Reviving traditional enclosures to encourage vegetation regeneration and reduce land degradation (*Ministry of Natural Resources and Tourism, Tanzania*)	MALI: Teaching farmers to collect climate data and integrate it into their planting decisions (*Government of Mali/Swiss Agency for Development and Cooperation*)	INDONESIA: Managing coral reefs in response to widespread coral bleaching (*WWF*)

On one end of the continuum, the '**vulnerability focus**' describes conventional development strategies, such as alleviating poverty, improving community health and health care, and increasing opportunities for education. Although these development strategies are not targeted at increasing preparedness or resilience against specific climate hazards, they can improve a community's capacity to react and adapt to a wide range of stresses brought on by climate change. Diversifying peoples' livelihoods, for instance, will likely reduce a community's vulnerability to a range of external shocks, whether climate related or not. This approach can be broadly understood as climate resilient development, in that the adaptation response is framed within a development perspective.

On the other end of the spectrum, the '**impact focus**' describes adaptation actions that are targeted at specific climate hazards, such as relocating an entire community or building coastal defenses in the face of sea level rise. These highly targeted, impact-focused 'discrete adaptation' actions are likely to become more necessary as climate change impacts increase. Impact-focused actions generally fall outside conventional development strategies: they aim to prevent catastrophic harm to the community, rather than to improve well-being.

Between these extremes lie strategies and actions that affirm, to varying degrees, both development and adaptation priorities, referred to as '**building adaptive capacity**' and '**managing climate risk**'. These approaches achieve development benefits while mitigating the risks of harm from climate impacts.

Traditional financial flows from aid (Official Development Assistance (ODA)) tend to focus mainly on the vulnerability end of the spectrum. Influenced by its focus on the additional human-caused element of climate change, UNFCCC sources of climate finance have tended to focus on the 'response to impacts' end of the spectrum.

Sources: McGray et al., 2007; Tanner and Mitchell, 2008.

Development and adaptation do not, however, always go hand in hand; although poverty is linked to vulnerability, the two are not synonymous. Unless carefully planned, poverty reduction and development efforts can actually heighten vulnerability to climate change. Activities that promote human settlement or infrastructure development in areas that may become unsuitable because of climate change, for example, only increase vulnerability. Development/poverty reduction efforts may also increase vulnerability for particular groups of people or adversely impact people at different spatial and temporal scales. In addition, although poor people's vulnerability is often used to justify adaptation, particularly through a social justice framing, many measures do not explicitly assess impacts on poor people or target the links between poverty and vulnerability (Eriksen and O'Brien, 2007).

3.3 How climate hazards translate into development impacts

Development experts and climate scientists typically understand climate risks and how to mitigate these risks differently. Climate scientists typically interpret climate risks as the probability of climate **hazards** occurring; adaptation is accordingly viewed in terms of the overall response of society or different sectors to these hazards, focusing on engineering approaches and structural measures (e.g. water diversion projects). Through a development lens, climate risks are understood as the probability of adverse human impacts associated with climate hazards, which is a function of the probability and magnitude of different impacts. As we will see, these in turn are mediated by vulnerability contexts.

Climate hazards, while globally distributed and blind to socio-economic circumstances, affect poor people disproportionately. This fact is starkly illustrated in Figure 2.6, which indicates that in any given year over the period 2000–2004, a person from a developing country was 79 times more likely to be afflicted by climate disasters than a person in a high-income country (UNDP, 2008). Moreover, there was a dramatic increase both in the incidence of climate risks and in the risk differential between rich and poor people to climate disasters over the past two decades. In other words, the climate dice are increasingly loaded against poor people.

A closer look at the main determinants of climate risks (defined as the probability of adverse human impacts) will help illuminate the

Figure 2.6 *Disaster risks are skewed towards developing countries.*

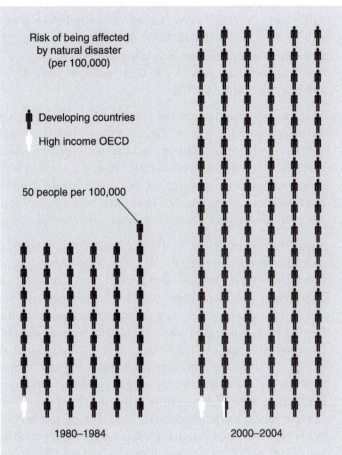

Source: HDRO calculations based on OFDA and CRED 2007 (UNDP, 2008).

asymmetric human impacts of climate events in different locations. There are four main components of climate risk: the probability that a climate hazard will occur (π); the degree of exposure to the hazard (E); the sensitivity to the hazard (S); and the adaptive capacities of the affected population (AC), (DFID, 2009). The relationship between these factors is represented as:

$$Climate\ risks = \pi * (E * S)/AC$$

Exposure, or proneness to climate hazards, relates to biophysical features such as geography, meteorology and geomorphology. As the IPCC assessments show, both temperature increases and precipitation changes will vary from region to region, affecting higher and lower latitudes differently. The tropics and sub-tropics are more likely to suffer from decreasing crop yields, communities in semi-arid areas will be most exposed to drought and soil erosion, coastal communities will be at higher risk from increased cyclone activity and sea level rise, and floods are expected to be a particular concern for temperate and humid regions (IPCC).

These biophysical factors and considerations alone do not account for the asymmetric impacts of climate change. They explain why the distribution of probabilities associated with specific climate hazards varies across space but reveal little about the varying scope and depth of impacts associated with a given climate hazard. Biophysical considerations alone thus fail to explain why similar climate hazards can produce a wide range of impacts in different locations.

Sensitivity is the degree to which climate hazards translate into shocks or stresses for any given community or ecosystem. It is a function of: (i) the resilience of critical infrastructure and ecosystems, (ii) the patterns of human settlements and activity, and (iii) the quality of climate defenses. Poor people tend to be located in marginal lands and depleted or stressed natural environments, and eke livelihoods out of resources that are sensitive to even small changes in climate conditions. Their infrastructure – i.e. their homes, water and energy supply systems, communication routes etc. – tends to be of low quality. Thus the lives, health and property of poor people are at greater risk from climate-related disasters. This helps explain why, even though Japan is more exposed to risks associated with cyclones and flooding than the Philippines, average fatalities between 2000 and 2004 were almost 11 times higher in the Philippines (711 to 66) (OFDA and CRED 2007, cited in UNDP, 2008).

The more affluent are better able to moderate their sensitivity to climate hazards. Their infrastructure is typically of higher quality, built to withstand a range of weather conditions and disaster scenarios in accordance with strict codes and standards. The Netherlands provides a good example of technological innovations and climate defense infrastructures that markedly reduce sensitivity to climate hazards. While most of the country is meters below sea level, the Dutch have suffered little from coastal storm surges thanks to their elaborate system of dykes, which act as a powerful buffer against these hazards.

Of course, individuals, communities, and societies are not passive in the face of climate shocks and stresses. Their **adaptive capacity** determines their ability to effectively respond to and moderate the impact of shock and stress, through anticipation, damage avoidance or mitigation, learning, coping and recovery. Adaptive capacity can been viewed as the means of securing the resources needed to reduce sensitivity, cope and bounce back after a shock.

While sensitivity relates primarily to technical aspects such as infrastructure and technology, adaptive capacity relates to broader socio-economic, cultural, institutional and political considerations such as income levels, educational attainment, the nature and strength of social networks, access to credit and insurance, governance, institutions, power relations and any other factor that affects the readiness and ability to deal with climate-related hazards.

Vulnerability to climate change can be understood as the combination of exposure, sensitivity and adaptive capacity, and is therefore determined by a range of biophysical, socio-economic, environmental, institutional and organizational factors (see also Chapter 5). While conventional development strategies may increase resilience, there are also cases of **maladaptation** when they can result in increased vulnerability to climate impacts.

The effect of economic growth on climate risks for example can work both ways. Economic growth can provide a form of insurance against climate risks and thereby increase resilience, by boosting a country's resources to withstand and recover from climate shocks. On the other hand, some growth strategies can add to non-economic dimensions of vulnerability, e.g. when they entail the large-scale depletion and degradation of the natural resource base on which the poor depend disproportionately for their subsistence and livelihoods, or when they undermine traditional and informal institutions on which the poor traditionally rely for social protection and assistance.

China provides a good example of where a stellar economic performance over three decades came at the cost of an escalating ecological crisis, mounting social disparities and endemic corruption. In all these areas the situation has worsened in tandem with economic reform, causing widespread discontent and threatening to jeopardize future growth (Horn and Warburton, 2007).

Poverty and low human development are both caused by and contribute to weak adaptive capacities and vulnerability. With meager resources to deal with external shocks and stresses, poor people are typically limited to a small set of hard choices, and driven to make painful trade-offs which may be detrimental to their health, security, well-being, recovery and future development prospects. For instance, to make up for losses poor families may have no other option but to cut down on household spending by sacrificing on food, health and education expenditures which typically make up the bulk of their household budget. This has been addressed in the literature and policy debate as depletion of human capital, which often goes along with the shedding of other assets (Fuentes and Seck, 2008). Thus poor people who live on the brink can easily be bumped into a downward spiral of chronic poverty at the slightest climate-related shock (see Box 2.5).

As an illustration of this, 'The State of Food Insecurity in the World Report 2009' (FAO/WFP) highlights how poor people have coped with food price rises by reducing dietary diversity and household expenditures on essentials such as education and health care. The impacts of food price increases and volatility can be particularly devastating for poor households, who according to FAO estimates spend between 60 and 80 per cent of their income on food (FAO, 2011).

Box 2.5

Climate risk and the poverty trap

The effect of asset shedding on a household's well-being can be long lasting and even permanent, as the depletion of human capital, productive and adaptive capacities locks people into self-reinforcing downward spirals of poverty and vulnerability. This desperate phenomenon is known as a '**poverty trap**'. Sometimes impacts are passed down generations, e.g. when under-nourishment of an infant leads to stunting and impairment of mental capabilities, or when educational attainment is curtailed because the warped logic of survival dictates that a child's labour on the farm is more valuable than his/her learning in a school (Fuentes and Seck, 2008).

The perception or expectation of a climate risk is sometimes enough in itself to set in motion sub-optimal livelihood strategies, as risk aversion translates into preferences for low-risk, low-return investments, e.g. in cropping choices, which keep people mired in poverty (UNDP, 2008). Thus, perversely, poor people's natural aversion to risk has the effect over time of increasing their vulnerability to risk factors.

Sources: Fuentes and Seck, 2008; Christiaensen and Dercon, 2007.

With the expected rise in the frequency and severity of climate hazards the devastation on poor people is likely to be cumulative, as each response to a shock reduces resilience to the next. Also, as climate extremes are 'covariant risks' (i.e. simultaneously affecting a wide range of people in a region), current safety nets – both formal (e.g. social assistance) and informal (e.g. social networks) – are likely to be overwhelmed. Thus poor countries and peoples' ability to cope with a changing climate is weakening, and this presents a major challenge for poverty reduction and human development (UNDP, 2008).

4. Bridging climate change and development responses

It will be clear from the foregoing that climate risks threaten to derail development, and conversely, development choices and the socio-economic pathways they give rise to are important determinants of climate outcomes. An added layer of complexity arises from the fact that climate actions are in themselves closely intertwined with development choices. Understanding how adaptation, mitigation and development efforts overlap and interact is key to developing effective, efficient and equitable solutions that advance climate protection and development goals.

Many development practitioners view change as linear, incremental and predictable, subject to certain conditions (or assumptions) holding. This is typified by most development agencies' use of the 'logical framework' (or 'logframe') as a tool for designing, planning, implementing and monitoring development interventions. A logframe spells out the desired development outcome(s) that the proposed intervention should help achieve, and defines project objectives, outputs and activities in a linear, cascading causal chain. It also identifies the risk factors that might disrupt the linear causal chain flowing upwards from activities to goal achievement. These risk factors are defined as external to and beyond the control of the intervention (otherwise they should be internalized, i.e. addressed within the project itself).

With climate change, this linear logic breaks down. Climate risks cannot be isolated and treated as exogenous risk factors because they are causally linked to development choices and pathways. The chains of mutual causation linking climate change and development are represented schematically in the infinity knot depicted in Figure 2.1.

As depicted in this figure, development choices affect climate risks both directly – since they shape vulnerability contexts, which mediate climate risks – and with a time lag, since development choices lead to different emissions pathways and hence climate outcomes.

Because development and climate outcomes are causally linked and co-determined by development choices and pathways, development and climate change challenges cannot be effectively tackled in isolation. Instead, integrative responses – climate responses that are sensitive to development needs and development efforts that are informed by climate-awareness – are required.

This section looks at how responses to climate change and development inter-relate, clarifying the need for integrative approaches. It then considers the barriers to and conditions for effective integration of climate concerns into development, and of development concerns into climate change responses.

4.1 Inter-relationships between development, adaptation and mitigation

Although development and climate change are clearly intertwined, they can be perceived as competing for financial resources and policy attention. Adaptation and mitigation measures and non-climate policies compete for scarce public and private resources at the global and (less so) national levels. The 2010 World Development Report notes policymakers' concerns about their development budgets being diverted to cope with weather-related emergencies (World Bank, 2010). While it is true that climate change investments can enhance the response capacities of future generations, they can also displace investments that might have created other opportunities for those same people in the future.

As the urgency of stabilizing the climate mounts, development choices will be increasingly influenced, if not shaped, by the pressing need to reduce emissions. The global mitigation imperative could limit 'development space' or provide new incentives and resources for clean development. In economies that are dependent on high-emitting productive sectors or fossil energy, emissions reduction is likely to impose transitory economic costs (although some industries such as clean technology and renewable energy industries may benefit directly).

Climate actions can also have potentially significant spillover effects that adversely affect development efforts. For example, increasing demand for biofuel crops can displace the production of staple food crops and contribute to food price increases that hurt the poor. We saw this dynamic play out in 2007–08 when US and EU biofuel mandates contributed to the global food price crisis. Furthermore, aggressive emissions-reduction policies that curb economic growth and slow development can result in greater vulnerability to climate risks by weakening countries' capacities to respond to these risks.

Development choices also affect climate responses and their effectiveness. This is particularly true with decisions relating to infrastructure, settlement and city design, which can create long-term commitments to wasteful patterns of fossil energy use (known as the **'lock-in' effect**), and result in maladaptation. For example, as a means of expanding access to energy, a development planner might favour building coal-fired power plants over developing renewable alternatives, on the basis of cost alone. But because power plants typically have life spans of 40–50 years, that decision would lock-in a higher-emissions energy supply system for decades to come. Likewise, settlement of coastal zones exposed to sea level rises will create acute vulnerability to climate change that will be extremely difficult and costly to reverse short of relocating entire communities. Unless carefully planned, development efforts can actually increase vulnerability to climate change.

There are also significant potential synergies in the short term between climate change and development responses. Specific examples of development co-benefits associated with adaptation and mitigation actions were pointed out earlier in this chapter and will be elaborated on in chapters 4 and 5.

The range and effectiveness of available adaptation and mitigation options is conditioned by the existence and nature of **adaptive and mitigative capacities**. These capacities reflect prevailing socio-economic, technological and environmental conditions, which are in turn shaped by development choices and pathways. Thus adaptive and mitigative capacities correlate with sustainable development indicators, such as income, education and ecosystem health. The pursuit of sustainable development contributes directly to boosting peoples' and societies' capacities to respond to climate change and thus to reducing climate risks/impacts (Najam *et al.*, 2007).

In addition to the co-benefits mentioned, climate actions can also help advance development efforts in a deeper sense. For example, mitigation measures that aim to correct **market failures** and distortions – by eliminating perverse fossil fuel subsidies that benefit industries and the rich at the expense of the poor, for example – advance sustainable development. Also, as we saw earlier, efforts to improve adaptive capacities also strengthen resilience to broader, non-climate shocks and stresses. Reducing vulnerability as a means of adapting to climate change therefore has a direct pay-off in terms of development outcomes.

Adaptation and mitigation linkages

Ultimately the degree of human impacts on climate change is determined by both adaptation and mitigation efforts. As we saw in section 3.2 of this chapter, there is no question that both adaptation and mitigation are essential to reduce the expected impacts of climate change. The absence of either would limit the effectiveness of the other as well as render its costs intolerably high. Over different time frames and geographical scales, there can be inconsistencies or trade-offs between the two, but there can also be potential synergies (Klein *et al.*, 2007).

On a global, long-term scale, adaptation and mitigation can be viewed as substitutes for each other: increased mitigation means less adaptation is necessary, and vice versa. On a shorter time scale, adaptation is the only option for reducing climate change impacts, since mitigation efforts take time to pay off and some warming is already unavoidable (*ibid*).

At the national and sub-national levels the adaptation–mitigation trade-off is less clear because effective mitigation depends on the combined efforts of a sufficient number of major greenhouse-gas emitters. Mitigation actions can, however, result in greater vulnerability to climate risks by weakening adaptive capacities. This may happen, for example, if aggressive emissions-reduction policies reduce resources available for adaptation and curb economic growth. Mitigation actions can also increase people's exposures to climate variability and hazards, such as when a shift to renewable energy sources like wind, hydro, and biofuels increases vulnerability to weather conditions.

In most sectors, the adaptation implications of mitigation actions and the contribution of adaptation actions to emissions are small. There are also a number of synergies between mitigation and adaptation efforts.

Mitigation efforts can foster adaptive capacity – and sustainable development – if they eliminate market failures and distortions, such as perverse subsidies, that prevent actors from making decisions on the basis of the true social costs of the available options. Creating synergies between adaptation and mitigation can result in more cost-effective actions. Opportunities for synergies can be significant in some sectors (e.g. agriculture and forestry, buildings and urban infrastructure) and limited in others (e.g. coastal systems, energy, health).

4.2 Towards integrated approaches to climate protection and development

It is now broadly recognized that climate actions and development decisions divorced from each other risk being ineffective and potentially counterproductive. Development choices can at times have a far greater influence on climate outcomes than mitigation actions, and climate risks can nullify development efforts if they are not accounted for development decisions. The IPCC Third Assessment Report concluded that the choice of development paths can lead to very different GHG emissions futures, and that charting a lower emissions development path will require major policy changes in areas other than climate change (Najam *et al.*, 2007). Addressing climate risks in development interventions will be vital, since up to 40 per cent of ODA is estimated to be sensitive to climate risks (World Bank, 2006).

This points to the need to integrate – or 'mainstream' – mitigation and adaptation concerns into development projects, policies and plans. **Mainstreaming** is particularly important because in most countries, climate matters remain under the purview of specialized agencies (e.g. environmental and **disaster risk reduction** authorities). We know, however, that the mix of policy instruments needed to combat climate change typically extends beyond what is in the toolkit of these specialized agencies. The deep inter-connections between economic and climate factors, the economy-wide nature of climate change impacts, and the profound economic implications of international climate agreements argue in favor of increased integration between climate change and economic policies.

Obstacles to bridging climate and development responses

Recognition of the need for and benefit of integrative approaches to tackling climate change and development is growing. Nevertheless, development and climate actions remain largely isolated from each other. This two-track approach can be attributed to several underlying factors:

- *Segmentation of knowledge* – Knowledge about climate change and development is constructed in very different ways. This variance creates a basic challenge to grappling with the nexus between the two. What we know about climate change stems from the work of physical and natural scientists, with an emphasis on positivist methods of analysis, including systems analysis and climate modelling. In contrast, development studies is the domain of the social sciences, involving a wide range of methods and disciplines, such as economics, sociology, philosophy, geography and anthropology. Likewise, research on adaptation and mitigation has been quite unconnected to date, involving largely different communities of scholars with different analytical approaches.
- *Institutional segmentation* – The parallel evolution of climate science and development studies within separate knowledge communities has long been reflected in decision-making: climate change and development challenges have historically been addressed separately as distinct and sometimes conflicting concerns. Institutional segmentation leads to adverse incentives for integration.
- *Decision makers lack awareness and understanding of climate–development linkages* – this is linked to points 1 and 2 above and may reflect a failure of the research community to communicate climate risks to decision makers in a language that they can easily understand and act upon.
- *Real or perceived conflicts between climate and development measures* – There can be real trade-offs between climate and development objectives. Sometimes these are merely perceived, and can be overcome with information. Trade-offs and synergies are not static, however. A relevant knowledge base enhances climate and development practitioners' ability to create synergies.
- *Limitation and segmentation of financial resources* – It can be difficult to make the case for investing in climate change adaptation and mitigation when urgent development challenges, such as hunger and poverty, also demand attention. Climate change investments often entail significant upfront costs for uncertain and distant benefits. Furthermore, the integration of climate change issues with other development policies and actions may be disincentivized by the creation of separate streams of climate finance.

Development and climate change responses typically involve different time horizons and entail different inter-generational distributions of costs and benefits. Responding to climate change requires a long-term commitment; the benefits often accrue to future generations while the costs are felt more immediately. Development is frequently measured by more immediate and tangible gains, for example, in GDP growth or the alleviation of extreme poverty.

Conditions for effective integration of climate and development responses

With the above in mind, the following approaches will be key to effectively integrating climate and development responses:

- *Generating decision-relevant information about climate–development links* – Decision makers need to be able to base their decisions on credible and context-specific information. Developing a strong evidence base around the links between climate change and development is a first step towards using this evidence consistently to embed climate change issues in development actions. Developing that evidence base will require integrated assessments and multi-disciplinary analysis of the interactions between climate change adaptation and mitigation, and development actions. To make use of the relevant analysis and evidence, decision makers need decision-support tools such as hazard mapping (for climate risks) and greenhouse gas emissions accounting.
- *Building technical and organizational capacities where they are needed* – In many developing countries, building the technical capacities to generate and interpret climate information will be a crucial precursor to effective bridging of climate and development responses. The design of climate policies and instrument mixes can be highly technical, and continuous monitoring, learning and adjustment will be necessary, particularly for more experimental measures.
- *Institutional design/reform* – An array of decisions must be made to influence emissions and development paths and vulnerability profiles. Coordinating policies in several sectors and at various scales will be a fundamental challenge. Social policies may be needed to compensate for the adverse social consequences that can result from some mitigation measures. Good institutional design allows relevant actors to coordinate and collectively prioritize and implement actions.

- *Public engagement and participation of relevant stakeholders in the making and implementation of decisions* – Given the uncertainties and long time horizons characteristic of adaptation decisions, effective public engagement is critical to ensure the legitimacy and durability of policy decisions. Public participation is also important to ensure that public values and interests are reflected in decisions about what constitutes acceptable levels of risk.
- *Adequate allocation of resources* – the important role of international climate finance is discussed in depth in Chapter 6. Beyond official sources of climate finance, agreeing a global price on carbon must be a priority in order to unleash private financing for climate action.

The literature reveals a strong consensus on the crucial roles of institutions and organizational capacities at various levels in shaping adaptation and mitigation responses. Furthermore, a significant body of research shows that democracies and nations with higher quality institutions suffer less death from natural disasters (Kahn, 2005; Ward and Shively, 2011). Sen's seminal work on hunger and famines sheds light on why democracies tend to be better not just at coping with disasters, but also at preventing them. Structures of accountability, buttressed by institutions such as the media, ensure that critical information flows both from the top down and from the bottom up to guide actions. Thus, underlying political structures have an important influence over the human costs of climate risks.

Besides institutions, the quality of policies and their execution count. Policy choices depend to a large degree on governmental capacities – technical, financial and political. Capacities to use, manage and act on climate information, and the degree to which relevant development policies incorporate considerations of and address climate risks are particularly relevant. Good policies can compensate for the shortcomings of markets and institutions by enhancing access and coverage of basic social services, and promoting poor people's direct participation in the decision-making process.

Conclusion

This chapter introduced basic concepts and analytical frameworks for investigating the relationship between climate change and development. It highlighted the fundamental inter-connectedness of climate and socio-economic systems and of actions to advance human development and protect the climate.

Equity, social justice and sustainability concerns are at the very core of how development and climate-protection actions are framed, justified and advocated. However, they remain insufficiently infused with the reality of development and climate policies and practices.

Stabilising the climate will require genuine collective action and the active participation of developing countries. This can entail significant upfront costs for uncertain and distant benefits. Yet mitigation is not necessarily a losing proposition for developing countries. The poorest people are often the most vulnerable to climate-related shocks and stresses; they therefore have the greatest stake in climate protection. Mitigation may also present significant development opportunities. Finally, developing countries may benefit from undertaking mitigation actions now in order to avoid retrofitting their economies if and when a global climate-policy **regime** requires significant emissions reductions in the future.

Even if aggressive action is taken to mitigate climate change, some degree of warming is already locked in to the climate system. Poor people stand to bear the brunt of climate impacts: they are typically highly exposed and sensitive to climate variations but have low adaptive capacity. In many cases development and adaptation actions are indistinguishable, but the two should not be viewed as synonymous. The need to adapt to climate change can provide an important impetus for development, but can also lead to policies that conflict with poverty reduction.

Not only is there room for 'win-win' strategies that reconcile development and climate responses, but each can only succeed in the long term if they are addressed as inter-connected challenges. To succeed, development and climate action will need to evolve in a way that emphasizes equity and sustainability. Section Two of this book addresses how climate change responses can be made more sensitive to development concerns. Section Three tackles the question of how future development efforts can be made compatible with climate protection.

Summary

- Poverty reduction and climate protection are interdependent challenges that share common characteristics. Notions of equity, social justice and sustainability provide common moral justifications and core principles guiding both.

● Effective mitigation requires developing countries to contribute to global emissions reductions. Rich countries will need to assist developing countries in transitioning to low-emissions development pathways, for instrumental as well as ethical reasons. Mitigation can also contribute in important ways to development in poor countries.

● Climate change adaptation will be necessary regardless of how effective global mitigation efforts are. Adaptation is particularly important for poor countries and people who are most vulnerable and will be hardest hit by the impacts of climate change.

● Understanding vulnerability contexts is key to understanding how climate hazards translate into development impacts.

● Understanding how adaptation, mitigation and development efforts overlap and interact is key to developing effective, efficient and equitable solutions that advance climate protection and development goals.

● Because development and climate outcomes are causally linked and co-determined by development choices and pathways, development and climate change challenges cannot be effectively tackled in isolation. Instead, integrative responses – climate responses that are sensitive to development needs and development efforts that are informed by climate-awareness – are required.

Discussion questions

● How do concerns of equity and sustainability shape development and climate action?

● Why should poor countries that contribute little to global greenhouse gas emissions undertake mitigation actions?

● How do vulnerability contexts mediate climate impacts on development?

● In what ways do climate actions contribute to or impede achievement of development goals?

- How can development and climate actions be made more complementary and mutually reinforcing?

Further reading

Adger, W.N., Huq, S., Brown, K., Conway, D. and Hulme, M. (2003) 'Adaptation to climate change in the developing world', *Progress in Development Studies*, 3.3: 179–195.

Grist, N. (2008) 'Positioning climate change in sustainable development discourse', *Journal of International Development*, 20.6: 783–803.

Huq, S., Reid, H. and Murray, L.A. (2006) *Climate Change and Development Links*, Gatekeeper series 123, London: IIED.

World Bank (2010) *World Development Report: Development and Climate Change*, Washington, D.C.: World Bank.

UNDP (2007) *Human Development Report 2007/2008, Fighting climate change: Human solidarity in a divided world*, New York: United Nations Development Programme.

Websites

http://cdkn.org/resources/
The Climate and Development Knowledge Network (CDKN) supports decision makers in designing and delivering climate compatible development through combining research, advisory services and knowledge management in support of locally owned and managed policy processes. The CDKN resource page contains a range of publications to aid decision makers in developing countries and those working in the field of climate compatible development.

www.oecd.org/dac/environment/climatechange
The OECD Climate Change and Development portal gives access to a wealth of research on three important aspects of the climate–development nexus: 1) climate change adaptation; 2) climate change finance effectiveness; and 3) measuring climate-related aid.

http://cait.wri.org
The World Resources Institute's Climate Analysis Indicators Tool (CAIT) provides comprehensive and comparable databases of greenhouse gas inventories and other climate-relevant data, analysis tools, and dynamic maps that can be used to analyze a wide range of climate-related data questions. CAIT's primary purpose is to provide high-quality information accessible through easy-to-use yet powerful and transparent analysis tools, to support the many dimensions of climate change policy.

http://climatedesk.org
The Climate Desk is a journalistic collaboration dedicated to exploring the impact – human, environmental, economic and political – of a changing climate. Its partners include some of the most respected names in media.

www.guardian.co.uk/global-development

The Guardian Global Development website is an excellent source of latest news on global development, and regularly features stories on the intersection of environment, climate and development. *The Guardian* is one of the UK's largest news organizations. This website is also a great resource for info-graphics, video resources illustrating important aspects of the climate–development nexus.

http://sdwebx.worldbank.org/climateportal/

World Bank Climate Change Knowledge Portal contains environmental, disaster risk, and socio-economic datasets, as well as synthesis products, such as the Climate Adaptation Country Profiles. The portal also provides intelligent links to other resources and tools.

http://country-profiles.geog.ox.ac.uk/#documentation

Climate Change Country Profiles

Joint initiative with DfID and the Oxford School of Geography and Environment to help address the climate change information gap for developing countries, by making use of existing climate data to generate a series of country-level studies of climate observations. Underlying data also available for use in further research. Each country profile contains a set of maps and diagrams demonstrating the observed and projected climates of that country. A narrative summarizes the data in the figures, and places it in the context of the country's general climate.

SECTION TWO
Responding to climate change as a development challenge

3 The international climate change regime

Based on consensual decision-making, the UN climate change negotiations can be a slow process.

- The UNFCCC and the international sustainable development context
- The architecture and evolution of the climate regime under the UNFCCC
- Key development aspects of the UNFCCC
- A framework for assessing the international response from a development perspective
- Thinking beyond the UNFCCC – other multi-lateral and transnational climate partnerships
- A simple development diagnostic for the international climate change response

Introduction

As outlined in Chapter 2, **climate change** represents a global collective action problem. Individual actors are likely to 'free-ride' on the actions of others unless a collective agreement is reached on respective actions. To date, the collective international response to climate change has been driven largely by the United Nations Framework Convention on Climate Change (**UNFCCC**). This chapter outlines the main features, evolution and future of the UNFCCC, focusing on the implications for and role of developing countries. We set out a number of key areas for the international response, designed to equip the reader to assess future responses from a development perspective. The chapter concludes by suggesting that the role of transnational networks and governance are becoming an increasingly important vehicle for global action on climate change, particularly in light of the failure to date of the UNFCCC to deliver actions at adequate speed and scale.

1. The UNFCCC and the international sustainable development context

The UNFCCC was established against the backdrop of growing international concern and evolving multi-lateral responses to global environmental problems. Following an international meeting of scientists at the First World Climate Conference in 1979, the international community began to recognize the scale of the climate change problem and the global nature of actions required to stabilize atmospheric **greenhouse gas** concentrations. The 1988 United Nations

General Assembly (UNGA) meeting endorsed a committee to negotiate an international response and the creation of the Intergovernmental Panel on Climate Change (IPCC) to provide a regular assessment of the available science that could inform policy and decision-making.

The resulting UNFCCC was launched as part of a wider set of responses to related global environmental problems at the World Conference on Environment and Development (the 'Rio Earth Summit') of 1992. The Rio Earth Summit has become almost synonymous with the concept of sustainable development. It represented a far reaching and comprehensive attempt to create international commitments, as well as the infrastructure to support their implementation, in the form of concrete international processes and mechanisms for enabling, measuring and tracking progress on a range of sustainable development issues.

Rather than treating the symptoms, developing-country negotiators and commentators emphasized the need to tackle the structural causes of environment problems that were inextricably linked to the inequities of the global economic **regime** and the position of developing countries within it (Adams, 2001; Najam, 2005). These included unequal trade rules, unsustainable levels of debt, unequal levels of consumption, and the economic structural adjustment programmes imposed on developing countries as conditions for international financing by institutions such as the World Bank and International Monetary Fund. By contrast, most of the rich industrialized countries had a narrower perspective, viewing it mainly as an environmental problem informed by the science (Dubash, 2009).

In the development of the Convention, the latter view was dominant: despite the radical approaches, the initial climate change response was largely taken as a pollution control issue, driven by scientific assessment, target-setting and technology (Depledge and Yamin, 2010). Nevertheless, over time the inter-relationships with development issues have grown stronger, and the binary distinction between industrialized and developing countries is now becoming less tenable. The development perspective has also driven the growing attention to **adaptation** to the unavoidable impacts of climate change (see Chapter 5) and pushed **mitigation** actions to

consider transitions to **low-emissions development** pathways
(see Chapter 4).

2. UNFCCC regime: Architecture and evolution

The UNFCCC is a broad framework agreement for action within which
more legally binding protocols can be developed and agreed through a
regular schedule of UNFCCC negotiations (known as the 'Conference
of Parties' or 'COP'). We refer here to the text of the UNFCCC
agreement itself as 'the Convention', while we use 'UNFCCC' to refer
to the wider set of international institutions and instruments of the
climate change regime, which provides the contractual environment for
collective action. Regimes are 'social institutions consisting of agreed-
upon principles, norms, rules, procedures, and programs that govern
the interactions of actors in specific issue areas' (Levy *et al.*, 1995:
274). While the focus of much analysis has been on the setting of
specific targets, the UNFCCC as a regime provides a forum for

Plate 3.1 *Plenary session of the COP16 meeting in Cancun, Mexico, 2010.*

Credit: WRI.

negotiation and deliberation with stakeholder groups, driving supporting policy, regulatory and legislative change at the national level. Crucially, it also provides long-term signals to institutions and actors at the national and sub-national level, who in turn can align their activities to the international effort and hold the national government to account for their international commitments (Depledge and Yamin, 2010).

2.1 Actors and governance

The UNFCCC system involves 196 country 'Parties', or nation-states. Each country has an assigned national 'focal point' for representation and negotiations, most commonly located within their ministry of environment. Negotiation meetings are also attended by non-governmental and inter-governmental organizations such as the United Nations agencies and the World Bank. Countries tend to negotiate as part of wider groups (see Box 3.1). Although the formal larger groups are increasingly being broken up by the changing geography of emissions, economic power and national interests, the Convention divides country Parties into three overlapping groups:

- Annex 1 Parties: Comprised of countries of the Organisation for Economic Cooperation and Development (OECD) (with the exception of Mexico and South Korea, who joined after 1992) and countries undergoing the process of transition to a market economy (economies in transition, or EITs), including Russia and many Eastern European countries.
- Annex 2 Parties: Comprised of only the countries of the OECD; relevant in commitments for international cooperation.
- Non-Annex 1 Parties: The majority of the world's countries, including newly industrialized nations such as Singapore and South Korea, the larger economies such as India, China and Brazil, oil-producing states such as Saudi Arabia and Venezuela, Small Island Developing States (SIDS) and the low-income countries of the world.

The UNFCCC is governed by the Conference of the Parties (COP), which normally meets once a year and is made up of representatives of countries that have signed the Convention. The formally agreed

Box 3.1

Country grouping under the UNFCCC

Countries with similar interests and viewpoints tend to negotiate in groups, saving negotiating time, although these groups are dynamic and new alliances form as issues and debates evolve.

Main developing countries ('non-Annex I') negotiating groups:

- *AOSIS*: The Alliance of Small Island States and low-lying countries sharing similar developmental and environmental concerns. AOSIS has a membership of 42 States and observers

- *LDCs*: The group of 48 Least Developed Countries with lowest socio-economic development and economic vulnerability in the world

- *African Group*: 53 members from the African continent

- *G-77 and China*: This group has 132 members and the chairmanship rotates on a regional basis (between Africa, Asia and Latin America and the Caribbean). The chair will often speak for the whole group, but where the sub-members such as LDCs or AOSIS have different positions, they will speak separately

Main industrialized country ('Annex I') negotiating groups

- *Umbrella group*: Australia, Canada, Japan, New Zealand, Norway, the Russian Federation, Ukraine and the United States

- *EU*: The 28 countries of the European Union

A case of its own:

- *Environmental Integrity Group*: Switzerland + Mexico and South Korea (both OECD)

Source: UNFCCC, 1992.

decisions are usually prepared in 'subsidiary bodies', also made up of country negotiators.

2.2 Objective, principles and commitments

The Convention sets out the basis for the international response. Readers are encouraged to become familiar with the Convention text as a whole; this section briefly outlines some of the elements key to a development perspective.

Objective

The UNFCCC has been ratified by almost all the countries of the world. The ultimate objective of the Convention, which is shared with the Kyoto Protocol, is to achieve the

> stabilization of greenhouse gas concentrations in the atmosphere at a level that would prevent dangerous anthropogenic interference with the **climate** system. Such a level should be achieved within a time-frame sufficient to allow ecosystems to adapt naturally to climate change, to ensure that food production is not threatened and to enable economic development to proceed in a sustainable manner.
>
> (UNFCCC, 1992: article 2)

The links in this objective to food production and economic development are particularly important for developing countries, whose economies are typically dominated by agricultural and natural resource sectors and are therefore more sensitive to climate variations (see Chapter 2).

Principles

A set of principles underpin the UNFCCC, with three inter-related principles in the Convention being particularly important from a development perspective:

- Precautionary principle
- **Equity**/common but differentiated responsibilities
- The right to promote sustainable development

Article 3.1 of the Convention states that Parties should protect the climate system 'on the basis of equity and in accordance with their common but differentiated responsibilities and respective capabilities' (UNFCCC, 1992). It implies there is a common responsibility to act in the global interest to address climate change, but that some countries have a greater obligation and ability to do so than others. This underpins the view that developing countries, who have historically contributed less to global environmental problems and have more limited resources to respond, should take on more limited commitments and should be assisted in actions they do take by financial and technical assistance from richer countries. This principle is reinforced by Article 3.2, which recognises the specific needs and special circumstances of developing country Parties who will be hardest hit, both by climate change impacts and in fulfilling their Convention commitments.

Article 3.3 supports the precautionary principle, stating that a lack of scientific certainty should not be used as a reason to delay taking action to prevent environmental damage or disasters, especially if that damage might be serious or irreversible. It has most frequently been invoked by small island nations who fear that if we wait for absolute scientific certainty, this may threaten their very existence due to sea level rise. Article 3.4 provides for countries' right to promote sustainable development, noting that actions to tackle climate change should be appropriate to national development contexts. This is consistent with the promotion of a supportive and open international economic system laid out in Article 3.5, which discourages use of climate change measures to restrict international trade.

Commitments

The Convention contains sets of commitments relating to reporting on and implementing policies and measures to tackle climate change. Commitments are differentiated between countries in the three Annexes (see Table 3.1). The Convention also highlights countries

Table 3.1 Differentiated commitments under the UN Framework Convention on Climate Change (UNFCCC)

Key Commitments	Annex 1 (OECD countries* + Economies in Transition (EITs))	Annex 2 (OECD countries* only)	Non Annex 1 (G77+China)
Mitigation: Emissions reductions and enhancement of sinks	Specific (although non-legally binding) commitment to emissions reductions with the aim of returning their GHG emissions to 1990 levels by year 2000	Provide finance for full cost of Convention reporting and incremental cost of mitigation and adaptation measures in developing countries. Promote the development and transfer of environmentally friendly technologies to EITs and developing countries	Unspecific commitment to implement policies and measures to tackle climate change
	All Parties to submit reports on the actions they are taking to implement the Convention, known as *National Communications*		
Reporting: Inventories and National Communications	Must report on actions more frequently and submit yearly emissions inventories		Report actions and inventories less often. Least Developed Countries (LDCs) report on their emissions at their discretion
Other	Participate in climate research, systematic observation and information exchange Promote climate change education, training and public awareness		

* Reflecting OCED membership in 1992.

requiring attention due to **vulnerability** to the impacts of climate change (Article 4.8), especially in the Least Developed Countries (LDCs) (Article 4.9), and those countries whose economies are highly dependent on income generated from fossil fuels and associated products who may suffer from mitigation measures implemented as a result of the Convention (Article 4.8).

2.3 Extending commitments: The Kyoto Protocol, post-2012 and post-2015

There have been three pivotal milestones in the development of the extended climate regime under the UNFCCC. The first was the creation of the Kyoto Protocol in 1997 to set binding targets for the Annex 1 countries. The second was the COP13 meeting in Bali in 2007 that established the process for developing the follow-up to Kyoto after its initial commitment period ended in 2012. The third was the agreement of the Durban Platform in 2011 to develop a comprehensive international agreement by 2015 following the failure to meet the 2009 deadline for a post-2012 agreement in Copenhagen.

The Kyoto Protocol

Developments in climate science in the early 1990s led to increasing acknowledgement that the Convention's voluntary commitments (summarized in table 3.1) were insufficient to meet its overall objective. As a result, a new set of talks began on legally binding targets culminating in the Kyoto Protocol of 1997, which came into legal effect in 2005. Sharing the Convention's objective and core institutions, the Protocol created legally binding targets for Annex 1 countries for an average 5.2 per cent reduction in greenhouse gas emissions from 1990 levels during 2008–12 (the first commitment period). Targets for each country were individually negotiated: for example, the EU agreed to an overall average reduction of 8 per cent among its member states, USA to a 7 per cent cut, Japan and Canada a 6 per cent cut, while some countries such as Russia had to return to 1990 levels and others were permitted an increase due to national circumstances. The Protocol included a number of flexibility mechanisms, including the Clean Development Mechanism (CDM), which enables investment in emissions-reducing actions in developing countries to generate credits towards Annex 1 country targets (see section 3.1 of this chapter).

Despite being central to the creation of the Protocol, following a change in government the USA decided in 2003 not to ratify the agreement, citing its negative impact on the domestic economy and international competitiveness. Australia refused to ratify until 2011, and Canada withdrew in 2009. The absence of these major emitting economies and the poor overall record on meeting targets has led some developing countries to lose trust in the developed countries (and to some extent in the process itself) (Roberts and Parks, 2007). It also calls into question the 'legally binding' nature of such international agreements.

From post-2012 to post-2015: Failure or a more comprehensive global deal?

The Bali Action Plan at the COP13 meeting in 2007 agreed to negotiate an agreement to follow the Kyoto Protocol's first commitment period ('post-2012') to conclude before the COP15 meeting of 2009 in Copenhagen. Two years of intense negotiations followed, combined with a growth in global public consciousness over the seriousness of the climate change issue and recognition of the explicit development implications of the climate change challenge. The negotiations were fraught with divisions and stalling tactics, with many Annex 1 countries anxious not to commit to legally binding emissions reductions unless the larger developing countries did likewise. Those same developing countries invoked equity and common but differentiated responsibilities to call for Annex 1 countries to take serious first steps towards commitments. The oil-producing OPEC countries continued their historic stalling tactics in the interests of protecting their own economies.

The result was that at the COP15 meeting of 2009 in Copenhagen, country Parties collectively failed to agree on a binding agreement for emissions reductions beyond 2012. The non-binding 'Copenhagen Accord' proved controversial as it was brokered by a small number of countries behind closed doors in the final hours. Nevertheless, at the COP16 meeting in Cancun in 2010, its contents were largely codified into formal UNFCCC decisions, including a process of pledging voluntary actions, open to all countries.

While the Cancun agreements failed to agree collective action at a scale consistent with agreed **stabilization** goals, they created a number of significant shifts in the regime (see Box 3.2). Countries and groups increasingly split from the single voice of the G77, including some African countries, and AOSIS and LDCs, who began to side more openly with progressive Annex 1 countries in pushing for deeper and quicker cuts to greenhouse gas emissions. These splits also reflect the influence of discussion fora outside the UNFCCC that included meetings of some LDCs and SIDS as the *Climate Vulnerable Forum*, and of around 40 'progressive' countries from across the Annexes under the *Cartagena Dialogue for Progressive Action* (Bausch and Mehling, 2011). These are discussed further in section 5 at the end of this chapter.

The 2011 UNFCCC COP17 meeting in Durban produced the 'Durban Platform', an agreement to finalize negotiations by 2015 on a comprehensive agreement to be in place by 2020. This was greeted with dismay by many activists as it pushed the previous 2012 deadline further into the future. However, it crucially agreed to include all countries in some form of agreement, possibly leading to a spectrum of different types and legal forms of commitments. Nevertheless, a Kyoto Protocol second commitment period was adopted in 2012. But without Canada, Japan, New Zealand, Russia or USA, emissions reductions commitments covered countries representing only 15% of global emissions.

Equity and ambition are the two key challenges for the international regime's evolution. The Durban COP of 2011 was the first to acknowledge the gap in 'ambition' between pledges and the emissions reductions required for preventing dangerous climate change. How to increase ambition to meet this gap is now central to discussions.

Rather than helping secure ambitious emissions reduction targets, the equity debate to date has centred largely on sharing the 'burden' of emissions reductions based on historical responsibility. Reframing the climate debate in a way that offers a broader, more holistic approach to equity, protects the interests of the most vulnerable, including women, and promotes a shift to low-carbon and climate resilient societies will be key to achieving agreement on an ambitious emissions reduction goal. Advocates from countries seeking ambitious action see this as involving consideration of 'respective capabilities' as well as 'common but differentiated responsibilities'. In other words, larger, richer developing countries are likely to be under pressure to commit to greater action based on their capabilities even where their historic responsibility for causing climate change is still much lower than many Annex 1 countries.

The counter argument remains that Annex 1 countries have not yet done enough to take their initial steps for emissions reductions. At the same time, small island and developing-country concerns will continue to ensure that equity and adaptation are linked, particularly in calling for significant flows of international financial support. Finally, there may be moves to link a centralized post-2015 agreement with more networked and bottom-up sets of actions to tackle climate change, something we return to in section 5 at the end of this chapter.

Box 3.2

From Copenhagen to 2015: A changing climate regime

The Copenhagen, Cancun and Durban meetings marked important changes to the climate regime, including:

- Illustrating the **delays and limitations** of a consensus and regime-based approach to solving the collective action problem presented by climate change.

- Raising climate change to the highest political level, with over 100 heads of state attending the Copenhagen meeting.

- For the first time, countries agreed an **overarching goal** of limiting global temperature rise to 2°C above pre-industrial levels and a peak in emissions 'as soon as possible', and acknowledged the more ambitious limit of 1.5°C demanded by the small islands group AOSIS.

- The potential for **different legal forms of agreement**, with Copenhagen starting a **'pledge and review' system** in which countries declare voluntary commitments and review them in light of the global situation. A post-2015 agreement may have more than one legal form of agreement.

- The emergence of **major emitting emerging economies**, namely Brazil, South Africa, India and China **(BASIC)** as **power brokers** in the international regime and the widening of country participation in emissions reductions through voluntary declarations on actions to tackle climate change.

- Agreement to some form of **monitoring, reporting and verification** (MRV) procedures for emissions reductions in all countries, something that had previously faced stiff resistance by developing countries.

- Scaling up of **long-term financing** for developing countries through establishment of an international Green Climate Fund, and commitment to provide up to US$30 billion of 'Fast Start' funding in the short term through existing institutions (see section 3.2 in this chapter, and Chapter 6).

- Providing a **balanced approach to mitigation and adaptation**, including promoting national adaptation plans and enhancing coordination through a UNFCCC Adaptation Committee, and raising the profile of land-use and forestry issues, including through REDD+ mechanisms.

- Creating a new **Technology Mechanism** to enhance action on technology development and transfer to support action on mitigation and adaptation, supported by a Technology Executive Committee and a Climate Technology Centre and Network.

3. Key UNFCCC mechanisms for developing country action

3.1 The Clean Development Mechanism (CDM)

While the Kyoto Protocol did not include emissions reductions targets for developing countries, through the Clean Development Mechanism (CDM) it included a means through which developing countries are able to receive financial assistance for taking actions to reduce greenhouse gases. Through the CDM, private investors in projects that generate emissions reductions receive credits (known as 'certified emissions reductions' or CERs), which can then be purchased by Annex 1 countries and contribute in part towards their domestic emissions reductions targets under the Protocol.

The CDM has been regarded with optimism, having the potential to simultaneously stimulate private sector investment and technology transfer, incentivize emissions reductions, foster sustainable development in developing countries, allow Kyoto targets to be achieved more cost-effectively, and bridge the Annex 1/non-Annex 1 divide in the UNFCCC (Matsuo, 2003).

Despite this potential, however, CDM has been critiqued from a range of perspectives. From a development perspective, CDM has largely failed to deliver broad-based investments and socio-economic benefits for poor countries (Boyd et al., 2007; Disch, 2010). This is in part because socio-economic objectives are poorly defined, with each country making its own definition of what constitutes 'sustainable development' benefits (Boyd et al., 2007). In addition, CDM projects have been highly concentrated in large-scale industrial projects covering HFC23 and N_2O gases, where development co-benefits are less likely.

CDM projects are also geographically concentrated in a small handful of emerging economies (especially India, China, Brazil and Mexico). CDM investment has been particularly slow in Africa, with less than 3 per cent of global projects in 2012. Only 0.2 per cent of CERs are expected to come from LDCs (De Lopez et al., 2009; www.cdmpipeline.org for latest figures).

To some extent, this is simply a legacy of private sector investment following the lowest-cost project opportunities and once these are exhausted or the market price for carbon credits increases, CDM projects will spread to higher cost options in a wider set of sectors and countries.

The CDM has also been criticized from an ethical standpoint for taking the pressure off rich countries to make domestic emissions reductions domestically, instead allowing them to buy their way out of taking local action. More radical critiques have called CDM 'carbon colonialism' in its use of developing countries as sinks for **GHG** emissions from industrialized nations (Bulkeley and Newell, 2010). From an emissions perspective, the CDM has been challenged over the **additionality** of its resulting emissions reductions due to the potential for including projects that may have happened regardless of the CDM or incentivizing highly polluting activities that could then generate revenues from CDM credits later on (Müller, 2009).

Internationally, the Gold Standard system emerged as an NGO-led response to these problems. The Gold Standard accredits CDM and other carbon offset projects that demonstrate real and permanent greenhouse gas reductions and sustainable development benefits in local communities that are measured, reported and verified (see www. cdmgoldstandard.org). Accredited projects are predominantly in renewable energy projects, where development benefits have been shown to be more commonplace (Drupp, 2011) and resulting CERs can command a higher market price. As of March 2013, the Gold Standard has certified more than 750 projects in over 60 countries. There are also national examples of development-friendly CDM practice, such as in Peru, where the CDM process prioritizes stakeholder involvement and undertakes on-site visits to confirm the projects' contribution to sustainable development, rather than only a desk-based assessment (Disch, 2010).

3.2 Reducing Emissions from Deforestation and forest Degradation (REDD+)

The need to engage with land use, land-use change and forestry (LULUCF) in emissions reductions and as a greenhouse gas sink has led to the emergence of initiatives under the banner of *Reducing Emissions from Deforestation and forest Degradation* (REDD+) (Angelsen *et al.*, 2012). REDD+ initiatives provide a means through which individuals, projects and communities in developing countries can be financially rewarded for reducing emissions from deforestation, forest degradation and enhancement of carbon stock. REDD+ has been regarded with optimism as an opportunity for many developing countries to participate in emissions reductions related to the forestry sector, generate economic and social benefits, and provide a potentially cost-effective mitigation option (Parker and Mitchell, 2009).

A range of REDD+ initiatives have emerged that trial the approach and prepare countries for participation in an international REDD+ scheme. These include the UN-REDD+ programme of the United Nations agencies, the World Bank's Forest Investment Program (FIP) and Forest Carbon Partnership Facility, and Norway's support for REDD+ preparations and activities in Brazil, Democratic Republic of Congo, Guyana, Indonesia and Tanzania. However, the lack of country-level experience or institutions combined with the absence of a global scale funding mechanisms mean that REDD+ remains in its formative stage (see Box 3.3).

Box 3.3

Key challenges for REDD+ development

REDD+ has been slow to operationalize because of a range of issues, including:

- Differences in the ideological basis for REDD+ mechanisms, including those stressing market efficiency, ecological integrity, social benefits, and institutional governance (Hiraldo and Tanner, 2011);

- How or whether to include safeguards to ensure co-benefits, including socio-economic benefits, the rights of indigenous people, and biodiversity conservation objectives;

- The use of market-based mechanisms versus use of funds to finance REDD+ (reflecting the wider debate in climate finance);

- Difficulty agreeing common approaches for establishing baselines and subsequent monitoring, reporting and verification (MRV) mechanisms to measure progress;

- Difficulty establishing country systems for these MRV issues;

- Whether the scope of REDD+ should be widened to include other land uses (known as REDD++), including the conservation of carbon locked away in soils, and agricultural practices, especially linked to 'Climate Smart Agriculture' (FAO, 2010).

Nevertheless, despite these issues, the REDD+ agenda has rapidly gained momentum. Crucially, REDD+ could use a hybrid model that includes both market-based mechanisms and voluntary donations. Future funding may therefore include direct carbon market funding (where buyers pay sellers for an environmental service such as REDD+), market-linked funding (where revenues are generated from auctions of emissions), voluntary funds, and debt-for-nature swaps (Parker and Mitchell, 2009). However this develops, REDD+ is firmly embedded in negotiations around the future global climate regime and will feed into the development of a Green Climate Fund sanctioned by the Cancun agreements of 2011 (UNFCCC, 2011).

Plate 3.2 *Deforested land in Indonesia.*

Credit: WRI / Beth Gingold.

Note: REDD+ initiatives provide a means through which individuals, projects and communities in developing countries can be financially rewarded for reducing emissions from deforestation, forest degradation and enhancement of carbon stock.

3.3 UNFCCC climate funds

As well as the wide range of bilateral and multi-lateral sources of finance for climate change activities (see Chapter 6), a number of international funds under the UNFCCC have also been established to assist developing countries in fulfilling their commitments to tackling climate change. These include:

- The **Special Climate Change Fund** (SCCF) to finance projects relating to adaptation, technology transfer and capacity building in a wide range of sectors including energy, transport, industry, agriculture, forestry and waste management, as well as economic diversification.
- The **Least Developed Countries Fund** (LDCF) was established to finance a work programme in LDCs, including the development and implementation of National Adaptation Programmes of Action (NAPAs, see section 3.4 of this chapter).
- The Kyoto Protocol **Adaptation Fund** (commonly known as the 'Adaptation Fund') to finance adaptation actions in developing countries.

Both the SCCF and LDCF are financed from voluntary donations by Annex 1 countries and managed by the Washington-based Global Environment Facility (GEF), which also manages the funding to help countries meet their reporting requirements under the Convention (see section 3.4 in this chapter). The (Kyoto Protocol) Adaptation Fund, in contrast to the other Convention funds, is financed primarily from a levy of 2 per cent on trading of CDM credits. Although revenues vary depending on CDM market trading, it is regarded as a more predictable source than traditional aid financing, and with greater developing country representation on its management board.

Recognizing that these mechanisms were not providing finance at a scale commensurate with need, the Cancun agreements of 2010 agreed the establishment of a Green Climate Fund to scale up long-term financing for developing countries, with a target of US$100 billion per year by 2020. It also called for Annex 1 countries to mobilize new and additional 'fast start' finance of roughly US$30 billion during the establishment of the governance, rules and procedures for the Green Climate Fund (then assumed to be 2010–12).

The Fund's design was undertaken by a Transitional Committee made up of UNFCCC country representatives, with the intention of becoming operational by 2014 (UNFCCC, 2011). The GCF is governed by a Board under the guidance of the COP. This Board follows the Adaptation Fund example and has equitable representation of developing countries and dedicated seats for the Least Developed Countries and Small Island Developing States, as well as two civil society and two private sector representatives as active observers. The requirement for a gender-sensitive approach is explicitly recognized in all funding, with initial balanced funding streams, or 'windows', on mitigation and adaptation and the flexibility to add other windows around REDD+, technology transfer or a small-grants facility (Schalatek et al., 2012).

The Green Climate Fund's modalities for disbursement will be a combination of grants, concessional loans and other financial instruments (UNFCCC, 2011). Like the Adaptation Fund, the GCF will operate on a 'direct access' model, allowing recipient countries direct access through accredited national and sub-national implementing entities who will be subject to best-practice standards for financial management. A private finance facility will also be established to enable direct and indirect funding access by the private sector, subject to national government approval. Nevertheless, the GCFs sources of long-term finance remain unclear at the time of writing, as is its relationship with the World Bank's Climate Investment Funds.

3.4 National planning and reporting

The UNFCCC has also developed reporting and action planning processes at the national level for technology, adaptation, and mitigation issues. In their *National Communications* submitted to the UNFCCC, non-Annex 1 countries each prepare a national inventory of anthropogenic emissions by sources and removal by sinks, the actions taken or envisaged to implement the Convention. It can also include a description of national and regional development priorities, objectives and circumstances, such as vulnerability profiles and emissions trends. *Technology Needs Assessments* (TNAs) have also been developed to communicate country-specific technology needs to the international community and feed into the other reporting processes. Since 2007, this basic reporting process has been supplemented by **Nationally Appropriate Mitigation Actions (NAMAs)**, which link national mitigation commitments with sustainable development, technology, financing and capacity-building needs. As the name suggests, NAMAs recognize that countries may take different actions according to their circumstances (for example energy efficiency or renewable energy options, or access to renewable energy technology).

Recognizing that the adaptation rather than the mitigation agenda was the most pressing concern in the Least Developed Countries, the LDC Work Programme under the UNFCCC has focused primarily on the preparation of National Adaptation Programmes of Action (NAPAs). NAPAs allow countries to identify priority activities that respond to their 'urgent and immediate' adaptation needs – those where delay in implementation would increase vulnerability or future costs. As well as indicating areas for international funding, the NAPA process also provides an initial cross-sectoral and cross-government basis for integrating adaptation into wider planning processes. Priority sectors are usually agriculture, forestry and fisheries, reflecting their importance in underpinning livelihoods, followed by water supplies, extreme events, and capacity building (UNFCCC, 2007). National Adaptation Plans (NAPs) are now following up the NAPA baseline by considering medium and long-term adaptation needs in LDCs and other countries, with guidelines for their completion issued by the UNFCCC's LDC Expert Group in 2013 (LEG, 2013).

3.5 Technology development and transfer

The Convention and Kyoto Protocol both include commitments from industrialized countries to facilitate the development and transfer of environmentally friendly technologies and low carbon technologies to EITs and developing countries. The Cancun Agreements of 2010 established a Technology Mechanism that combines a Technology Executive Committee of experts with a Climate Technology Centre to facilitate an international technology network. It also acknowledged earlier calls for development and transfer of technologies for adaptation, such as new agricultural techniques or seed varieties (UNFCCC, 2006).

The role of technology is often regarded as central to tackling the climate change challenge, particularly for enabling transitions to lower carbon sources of energy. Two key debates have dominated the climate technology issue (see Figure 3.1). The first refers to the nature of the technology innovation process itself (Grubb, 2004). This is split between a 'technology push' view, which calls for large scale, publicly-funded research and development to provide the new technologies

Figure 3.1 *Dimensions of technology innovation and transfer.*

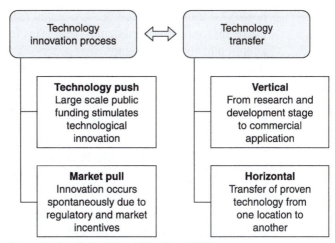

Source: Based on Grubb, 2004 and Ockwell *et al.*, 2008.

required to tackle climate change, and a 'market pull' view that sees technological innovation occurring spontaneously in the private sector as a response to regulatory and market incentives. Proponents of the former argue that climate change presents a new challenge in requiring an accelerated process of climate-related technology development that may be beyond the speed of market pull alone, requiring innovation oriented at public rather than private ownership.

Overlaid onto this is the issue of 'technology transfer'. This refers to both the 'vertical transfer' of technology from the research and development stage through to commercial application, while horizontal technology transfer is used to describe its movement from one location to another (Ockwell *et al.*, 2008). Debate in the UNFCCC has been dominated by the latter: transferring technology owned by businesses in the industrialized countries to the EITs and developing countries. This has led to negotiations dominated by discussions over how to secure access to technology hardware for developing countries given the trade barriers, patents and intellectual property rights (Ockwell *et al.*, 2008).

The debate has evolved in recent years. The global picture has become more complex than a simple Annex 1 country to non-Annex 1 country transfer, especially given the increasingly dominant role played by the larger emerging economies such as Brazil, India and China in developing and manufacturing of low carbon technologies. At the same time, technology issues in the UNFCCC increasingly stress how international cooperation can facilitate vertical rather than horizontal transfer, reflecting the fact that many technologies are not yet ready for commercial deployment. This has entailed greater attention to supporting innovation processes and in-country technology development. UNFCCC technology issues therefore now increasingly emphasize the provision of financial resources, collaborative research and development, and building national and regional capacity. This move to vertical transfer issues 'introduces a whole range of additional considerations, such as how to overcome high levels of investor risk for technologies at early stages of development, or how to adapt technologies to new contexts – whether economic, environmental or socio-technical' (Ockwell and Mallett, 2012: 7).

From a development perspective, the limited number of technologies and countries being funded through the UNFCCC mechanisms has raised questions about the extent to which poor people or poor countries are benefiting. China and India have received 80 per cent of funds to date and over 75 per cent of registered CDM projects are implemented using just five types of technology – hydro, methane avoidance, wind, biomass energy and landfill gas (Byrne *et al.*, 2012). Drawing in these core development issues, including reflecting technology needs in adaptation as well as mitigation, will be crucial to a future climate regime that delivers on both development and climate-stabilization objectives.

3.6 Loss and damage: A critical emerging issue

An important emerging UNFCCC issue for developing countries is the negotiations on 'loss and damage' in particularly vulnerable developing countries. This refers to tackling the residual impacts of climate change for which adaptation is not sufficient. Small island 'AOSIS' countries have long called for a mechanism to compensate for loss and damage related to climate change impacts, but issues of liability and compensation have so for made it too controversial for most Annex 1 countries.

Recently, however, there has been more widespread acknowledgement of the inevitability of negative climate change impacts. In part this reflects the failure of mitigation efforts to avoid dangerous climate change and the 2°C threshold (UNEP, 2011; World Bank, 2012d)). It stems in part also from the growing scientific evidence that climate change impacts are already discernible, particularly in natural systems (IPCC, 2007) and the limits to adaptation (Adger *et al.*, 2009).

After pressure from AOSIS and the LDCs, a UNFCCC loss and damage work programme was established in 2010. The Warsaw COP of 2013 saw initial discussions on establishing an international mechanism to address loss and damage and the role of the Convention in:

(a) Enhancing knowledge and understanding of comprehensive risk management approaches;
(b) Strengthening dialogue, coordination, coherence and synergies among relevant stakeholders; and
(c) Enhancing action and support, including finance, technology and capacity-building (UNFCCC, 2012: 22).

Loss and damage faces a number of critical conceptual and operational challenges (Wrathall *et al.*, 2013). Firstly, it remains scientifically difficult to attribute individual weather events to climate change rather than natural climate variability. The climate change problem also relates not only to specific 'events' such as storms but also to dynamic, creeping changes such as sea level rise or ocean acidification. Secondly, climate impacts are determined not only by a changing climate, but also by the changing patterns of human vulnerability (IPCC, 2012). For example, enhanced flooding may be due to a complex combination of intense rainfall, deforestation and increased building on flood plains.

Thirdly, an approach to loss and damage that seeks primarily to restore and rehabilitate what has been damaged will not promote the transformation of the structures that produce vulnerability in the first place (see Chapter 8). There is also concern that an emphasis on compensation for losses does not reduce the emphasis or urgency of mitigation. There also remain issues around how to value loss and damaged assets such as lives or cultural assets, and in ensuring delivery of assistance to those who most warrant it (Wrathall *et al.*, 2013). In addition, assessing different levels of responsibility (and therefore who should pay) for assistance faces similar challenges to the issue of burden sharing outlined in section 4.2 later in this chapter.

Although a thorny issue, loss and damage is likely to become increasingly important. Recognizing climate change impacts that outstrip their adaptive capacity, vulnerable developing countries will seek to develop both their own capacity to assess and tackle loss and damage, and an international mechanism to assist their response. At the same time, perhaps the seriousness of the loss and damage agenda will sharpen awareness of the consequences of failing to avoid dangerous climate change, becoming a rallying point for more ambitious mitigation actions (Warner and Zakieldeen, 2012).

4. A framework for assessing the international response

Given the evolving nature of the international negotiations, this section guides the reader through the key issues with which to analyze future agreements, policies and measures from a development perspective. The chapter ends with a set of diagnostic questions with which to interrogate any future proposal for international climate change action. We break down any assessment in terms of the ability to deliver a response that is both:

- Equitable (a fair distribution of costs and benefits, and broad participation in decision making).
- Effective (able to deliver climate-stabilization goals).

4.1 A fair response: Equity and social justice

'Climate justice' is increasingly used by international development NGOs in their climate change campaigns to lobby for a fair climate regime, particularly with regards to poorer developing countries (see Box 3.4). The principle of equity enshrined in the Convention is underpinned by the idea of fairness and theories of **social justice**. It can be divided into aspects of:

- **Distributive equity**: Concerned with the way costs and benefits are distributed among people and groups with competing interests.
- **Procedural equity**: Concerned with the fairness of planning and decision-making processes, including ensuring voice and participation for affected and interested groups.

Distributional equity issues have been crucial to both determining and frustrating international agreement on climate change. The Convention and Kyoto Protocol differentiate commitments between countries based on a binary distinction between two groups of countries (Annex 1 and non-Annex 1), and the definition of targets is set through negotiation rather than according to objective formula. There have historically been contrasting views regarding equity:

Industrialized countries (Annex 1) regard equity primarily in terms of 'meaningful participation' of other countries to ensure fairness in sharing the costs of mitigation efforts, especially given the growing emissions contribution and international trade competition from large, rapidly industrializing countries such as Brazil, India and China. They argue that this is consistent with the polluter-pays principle underpinning the Convention.

Developing countries (Non-Annex 1) on the other hand regard equity in terms of the need for those with historic responsibility (Annex 1 countries) to take action first, consistent with the principle of common but differentiated responsibilities. Equity is then used to lobby for redistribution and resource transfers for both mitigation and adaptation on the basis that these countries contributed least historically to its causes and are likely to suffer most from its consequences.

With the emergence of growing middle-income economies, this binary distinction is increasingly being broken down for a number of reasons. Negotiations for a post-2020 climate agreement take place in a global context that is radically different from that when the Kyoto Protocol was agreed in 1997. The Protocol has certainly not delivered at the scale intended due to poor performance against targets and the withdrawal of USA and Canada. At the same time, developed countries no longer account for the majority of greenhouse gas emissions, with non-Annex 1 countries now contributing over half of CO_2 emissions and making up seven of the largest 15 emitters (see Figure 3.2). Seven non-Annex 1 countries have joined the OECD group of industrialized nations since 1997, while China, India, Brazil and South Africa have dramatically increased their international political and economic power, as well as their greenhouse gas emissions. China overtook the US as the major emitter of greenhouse gases in 2007, and Indonesia is the 3rd largest emitter due to deforestation (World Bank, 2010a).

Despite these changes, per capita distributions are generally still highest in Annex 1 countries and the historic and current contribution to global emissions by low-income countries remains small, even including the influence of land-use change (see Figure 3.2). As a consequence, the challenge for the future regime is to distinguish in particular between the higher historic emitters, the newly industrializing countries where emissions are growing rapidly, and the poorest developing countries, who have both little historic responsibility and little prospect of significant emissions growth in the near future.

Figure 3.2 *High-income countries have historically contributed a disproportionate share of global emissions and still do.*

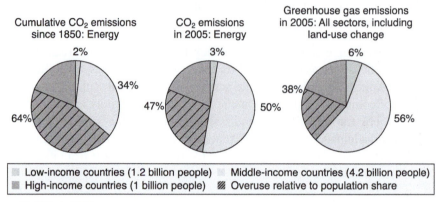

Share of global emissions, historic and 2005

Cumulative CO$_2$ emissions since 1850: Energy

CO$_2$ emissions in 2005: Energy

Greenhouse gas emissions in 2005: All sectors, including land-use change

2% 34% 64%

3% 47% 50%

6% 38% 56%

Low-income countries (1.2 billion people) Middle-income countries (4.2 billion people)
High-income countries (1 billion people) Overuse relative to population share

Source: World Bank, 2010a.

Box 3.4

Defining climate justice

Notions of social justice and equity have been central to many of the global climate change campaigns driven by pressure groups, especially those with an international development dimension. The *Mary Robinson Foundation – Climate Justice* (MRFCJ) was founded by former President of Ireland Mary Robinson to pursue a climate justice agenda based on the need to:

● Respect and protect human rights

● Support the right to development

● Share benefits and burdens equitably

● Ensure that decisions on climate change are participatory, transparent and accountable

● Highlight gender equality and equity

● Harness the transformative power of education for climate stewardship

● Use effective partnerships to secure climate justice

Source: www.mrfcj.org/about

4.2 Distributional equity: Burden-sharing and financing adaptation

Principles of distributional equity have been applied in particular to the allocation of emissions reductions between countries. In the climate-policy jargon, this is known as 'burden-sharing' and can be divided into five broad categories of proposed options (see Table 3.2).

Table 3.2 *Main approaches to burden-sharing of global emissions reductions*

Approach	Characteristics	Proposals / Application	Distributive Equity issues
Grandfathering	Emissions reductions targets as a percentage of emissions in a baseline year	Kyoto protocol based on this approach: Total 6–7% reduction below 1990 baseline by 2010	May be unfair if applied to developing countries, whose emissions are low but are rising rapidly
Carbon intensity targets	Reductions in emissions as a proportion of economic activity (Gross Domestic Product – GDP)	Proposed by USA following their withdrawal from Kyoto. Applied in China's pledge to reduce carbon intensity by 40–45% from 2005 levels by 2020 as a pledge following the Copenhagen summit	Does not impose a 'hard cap' on developing countries' growth, but it may be harder to reduce intensity for countries at early stages of industrial development
Historical responsibility	Based on polluter-pays principle and 'ecological debt'. Recognizes that CO_2 remains in atmosphere for 100–120 years	'Brazilian proposal' of early 2000s. Resisted by some industrialized countries who argued that historic emissions were made without an understanding of their damaging effects	'Early industrialising' countries make majority of cuts, recognizing damage caused by cumulative emissions.
Per capita	All humans have equal rights to global atmosphere; targets based on a per capita division of an agreed total global emissions budget	'Contraction and convergence' approach in which richer countries contract and all countries converge on common per capita emissions and then contract together. Also reflected in the Brazilian proposal	Industrialised and oil-intensive economies required to do more as most already above the per capita emissions average. Creates 'carbon space' for developing countries to grow their economies initially
Hybrid approaches	Combine aspects of responsibility in above approaches with the ability to take action or pay for international cooperation	Pew Centre proposal Triptych approach Greenhouse Rights Development Framework (GRDs) (see Box 3.5)	As above but also reflects capacity to take action and equity within as well as between countries

Source: adapted from Roberts and Parks, 2007.

Per capita allocation based on population has been central to many development and equity-focused approaches to burden-sharing. These assign emissions allocations for each citizen on the planet based on an overall emissions stabilization target, usually including a target for when emissions will peak, followed by a joint reduction in emissions. As most developing countries have total emissions significantly below the combined total allocations of their population, this allocation recognizes **carbon space** as a right to development to grow economies and meet basic needs.

The per capita allocation approach was notably pioneered by the 'Contraction and Convergence' approach originally developed by Aubrey Mayer at the Global Commons Institute. It calls for stringent and early cuts to industrialized country emissions and allows for the growth of emissions in developing countries up until a global emissions peak at which per capita emissions are equal across all countries of the world (the period of 'convergence'). Per capita emissions then decline together at the same rate ('contraction') to meet an agreed stabilization target, with different curve shapes depending on that target and time that convergence is reached (see Figure 3.3).

The Greenhouse Rights Development Framework (GRDF) linked a population-based emissions allocation with a GDP-based proxy for the ability to pay for actions ('respective capabilities' in UNFCCC language) (Baer *et al.*, 2009; see Box 3.5). Under this proposal, almost all countries of the world have some obligations to pay for climate policies (both mitigation and adaptation) on the basis of capacity (ability to pay) and responsibility (contribution to the problem). This idea is central to breaking down the binary divide between 'rich' and 'poor' countries when allocating commitments in the international regime.

The GRDF incorporates distributional equity related to impacts and adaptation issues. Following polluter-pays principles, a fair outcome would see richer nations (and richer citizens) compensating poorer nations (and citizens) for the damages from anthropogenic climate change impacts caused by greenhouse gas emissions. Although the science of separating climate change signals from regular climate variability makes a perfect calculation of such payments difficult, this equity principle underpins the provision of adaptation funding under the UNFCCC (Adger *et al.*, 2006; see Chapter 6).

Figure 3.3 *The Contraction and Convergence approach.*

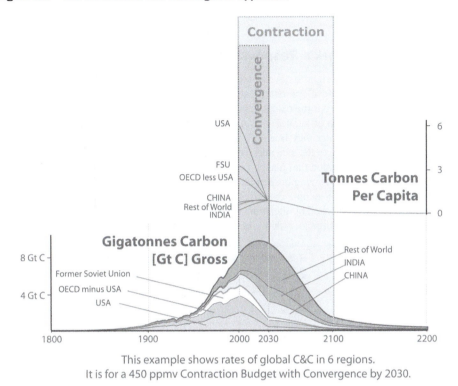

This example shows rates of global C&C in 6 regions.
It is for a 450 ppmv Contraction Budget with Convergence by 2030.

Source: GCI, 2012.

4.3 Procedural equity: Voice, recognition and participation

Whilst distributional issues tend to dominate in discussions of equity, a development-based appraisal of the international response must also consider procedural justice: how the interests, concerns and perspectives of all countries and citizens are considered and reflected in the decision-making process. On the one hand, procedural equity is supported by the UNFCCC system of one-country, one-vote and consensus decision-making. This means that, in principle, even the smallest country is able to have a say and veto decisions. At the same time, smaller developing countries are constrained by the size, technical and analytical capacity of their delegations, including preparatory activities, as well as their limited economic and diplomatic power in the global sphere. As an example, the dominance of the mitigation agenda over the adaptation agenda during the early years of the UNFCCC

Box 3.5

The GRD framework's Responsibility and Capacity Index

The GRDF embodies a 'right to development' by excluding the income and emissions of individuals below a 'development threshold' (set at $7,500 per capita, purchasing power parity adjusted) from the calculation of responsibility and capacity. In doing so, it examines the distribution of income *within* countries and treats people of equal wealth similarly, whatever country they live in. Thus even poor countries have some obligations, proportional to the size and wealth of their middle and upper classes. The result is a global distribution that places very limited obligations on low-income countries, a limited but growing burden on middle-income countries and a large but declining burden on high-income countries. This demonstrates the considerable differentiation that is masked by the Annex 1/Non-Annex 1 groupings under the UNFCCC, which groups the low- and middle-income countries together.

	Responsibility and Capacity Index (% of global total)	
	2020	2030
High-income countries	69%	61%
Middle-income countries	30%	38%
Low-income countries	0.3%	0.5%

Source: Baer *et al.*, 2009.

reflects the differences in recognition and voice. Developing country participation in the climate regime was therefore largely limited to their role in mitigation, rather than participation in the terms of their own pressing priorities of adapting to climate change impacts.

Deliberative justice can also be considered from the perspective of gender equality (see Box 3.6) and ensuring a diverse range of non-governmental voices are able to influence both the international debate and national government views. This will depend on both the nature of the international regime and spaces for deliberation in individual countries. At the international level, the UNFCCC makes provision for a range of different non-governmental coalitions ('constituencies') as observers to the negotiations (see Box 3.7), alongside other coalitions of actors lobbying around specific agendas such as forestry, nuclear or solar energy. These groups employ different tactics to influence agenda-setting and decision-making, including advocacy campaigns, research studies and personal lobbying of government representatives. Their capacity to do so is again heavily influenced by their access to resources.

Box 3.6

Participation and gender considerations in the UNFCCC

Gender considerations and representation have been sorely lacking from the international climate change negotiations. Although female representation in COP delegations has been low (as low as 15 per cent at COP3) women have played a key role in the negotiations; female leaders of the German and Swiss negotiating teams in 1997 have been praised for proactively linking with developing countries, helping their much smaller teams integrate into the negotiations. Likewise, female negotiators from developing countries including Peru and Zimbabwe were noted for their unifying role. Although female representation does not guarantee that gender issues will be properly taken into account, and conversely men may bring strong gender perspectives to the negotiations, the evidence suggests that key male negotiators are not making gender considerations enough of a priority, and that the best way to address this is to have more female representatives involved in the negotiations. However, given the current disproportional male representation in this area, it is vital that men are aware of and actively champion gender issues, rather than wait for gender equality in representation if gender issues are to be properly addressed at the negotiating table.

Source: Haigh and Vallely, 2010.

Box 3.7

UNFCCC observer group 'constituencies'

- Business and industry non-governmental organizations (BINGO)
- Environmental non-governmental organizations (ENGO)
- Research and independent non-governmental organizations (RINGO)
- Local governments and municipal authorities (LGMA)
- Indigenous peoples organizations (IPO)
- Youth non-governmental organizations (YOUNGO)
- Trade union non-governmental organizations (TUNGO)
- Women and gender organizations

Plate 3.3 *Protests during the COP 17 meeting in Durban. At the annual UNFCCC Conference of the Parties (COP) meetings, NGO observer groups use a mix of policy proposals, advocacy and direct action to lobby the country delegates.*

Credit: WRI /Michael Oko.

At a wider level, global inequalities are reinforced by differences in the availability of data, analysis and the capacity to advance scientific understanding around the world. For example, climate data records are generally longer and better quality in the more industrialized nations, with large gaps in historical data in some parts of the world, especially in sub-Saharan Africa. At the same time, climate change expertise is also unevenly distributed and unequally represented in global assessment processes such as the IPCC, even when differences in capacity, population and economic development are taken into account (Ho-Lem *et al.*, 2011).

4.4 Climate stabilization: The ultimate equity goal

From a development standpoint, the international response must ultimately be assessed in terms of effectiveness in stabilizing the climate at safe levels. The lack of clarity in the Convention has led to ongoing scientific and political debate over what level of atmospheric greenhouse gas concentrations constitutes 'dangerous anthropogenic interference'. As noted previously, while there is international agreement since the Copenhagen COP to aim to limit global average temperature rise to 2°C (and noting demands by the AOSIS group for a 1.5°C goal), this is not translated into consequent concentration levels in order to determine a timetable for global emissions reductions. In other words, without agreeing on the size of the pie, it is hard to allocate shares of it to different countries.

Given the significance of climate change impacts on developing countries and poor people's well-being, achieving stabilization at safe levels and thereby limiting the amount of adaptation required can be regarded as the most pressing development concern. However, as noted elsewhere in this book, in spite of the efforts of the UNFCCC and the Kyoto Protocol global greenhouse gases emissions have continued to rise and it is highly unlikely that temperatures will remain within the safety 'guardrail' of a 2°C temperature rise (Anderson and Bows, 2008; UNEP, 2011a). The cumulative impact of the post-Copenhagen national emissions reductions pledges is a long way short of this guardrail, with resulting global-mean temperature increase of 3.3°C above pre-industrial levels by 2100, lying within a range of 2.7°C and 4.2°C due to carbon-cycle and climate-modelling uncertainties (www.climateactiontracker.org; Höhne et al., 2012).

The ultimate equity goal could therefore be securing the pathway to a lower carbon future that prevents dangerous climate change and impacts that are much harder to adapt to. As such, common but differentiated responsibilities and respective capabilities, burden-sharing and adaptation assistance must be considered in the context of how to achieve stabilization, especially given the poor record to date of curbing emissions sufficiently. Nevertheless, assessments by the UN Environment Programme (UNEP, 2011a) and others have argued not only that urgent emissions reductions are required, they are also technically feasible to close the gap between the current emissions trajectory and that required for a likely chance of limiting global warming to 2°C (see Box 3.8).

Box 3.8

The 2°C target and the global emissions gap

Studies show that emission levels of approximately 44 gigatonnes of carbon dioxide equivalent ($GtCO_2e$) (range: 39–44 $GtCO_2e$*) in 2020 would be consistent with a 'likely' chance of limiting global warming to 2°C.

● Under business-as-usual projections, global emissions could reach 56 $GtCO_2e$ (range: 54–60 $GtCO_2e$) in 2020, leaving a gap of 12 $GtCO_2e$.

● If the lowest-ambition pledges were implemented in a 'lenient' rather than strict fashion, emissions could be lowered slightly to 53 $GtCO_2e$ (range: 52–57 $GtCO_2e$), leaving a significant gap of 9 $GtCO_2e$.

● The gap could be reduced substantially by policy options being discussed in the negotiations:

  ○ by countries moving to higher ambition, conditional pledges

  ○ by the negotiations adopting rules that avoid a net increase in emissions from:

    (a) 'lenient' accounting of land use, land-use change and forestry activities; and
    (b) the use of surplus emission units.

● If the above policy options were to be implemented, emissions in 2020 could be lowered to 49 $GtCO_2e$ (range: 47–51 $GtCO_2e$), reducing the size of the gap to 5 $GtCO_2e$. This is approximately equal to the annual global emissions from all the world's cars, buses and transport in 2005 – but this is also almost 60 per cent of the way towards reaching the 2°C target.

● Studies show that it is feasible to bridge the remaining gap through more ambitious domestic actions, some of which could be supported by international climate finance.

● With or without a gap, current studies indicate that steep emissions reductions are needed post-2020 in order to keep our chances of limiting warming to 2°C or 1.5°C.

Source: UNEP, 2011.

* Gigatons of CO_2 equivalent.

5. Beyond UNFCCC – other multi-lateral and transnational climate partnerships

There has been growing attention in recent years to the role of other partnerships outside the multi-lateral UNFCCC regime in driving the collective global response to climate change. One option has been to consider how to make progress through other multi-lateral institutions, many of which relate to climate change. Some have sought to re-emphasize the important role of multi-lateral solutions in tacking the multiple collective challenges facing the world (Evans, 2010). Others, such as Daniel Bodansky (2011), assess the potential effectiveness of other multi-lateral options not just in terms of the failures of the UNFCCC but also due to:

● The track record of success, trust and cooperation of related environmental regimes such as the Montreal Protocol on Substances that Deplete the Ozone Layer;
● The relative simplicity of dealing with single sectors and the tradition of cooperation among sectoral organizations such as the International Maritime Organisation (IMO) and International Civil Aviation Organisation (ICAO);
● The potential of regional fora for action, such as the Convention on Long-Range Trans-boundary Air Pollution (CLRTAP), which are based on established relations and may be appropriate for regional aspects of climate change.

There has also been a dramatic rise in transnational responses linking select countries, locations and actors. This has emerged in part from frustration at failure of the UNFCCC process to date to both regulate and stimulate implementation of emission reductions or adaptation assistance at levels commensurate with the problem (Bäckstrand, 2008; Bodansky, 2011). More positively, transnational responses can be seen as well suited to the climate change problem due to its complexity and the way it cuts across sectors and actors (Andonova et al., 2009). At the same time, globalization and the incredible advances in information and communication technology since the turn of the millennium have enabled previously unconnected (and sometimes un-connectable) groups of actors to form alliances, act jointly and learn from one another.

Andonova and colleagues (2009) identify three functional categories of transnational climate partnerships: information-sharing; capacity building and implementation; and rule-setting. To some extent transnational partnerships can be seen in terms of how they contribute to and shape international regimes, while 'others have considered the ways in which such actors have become "governors" in the environmental domain in their own right, beneath, around, and alongside interstate regime frameworks' (Bulkeley and Jordan, 2012: p. 556). This is reflected in national government partnerships such the Group of 8 (G8) forum of the world's wealthiest countries and its corollary the G8+5 (which adds China, India, Brazil, Mexico and South Africa), where discussions on climate change have mostly attempted to reinforce the UNFCCC. Others, such as the Asia-Pacific Partnership on Clean Development and Climate (APP) established by President Bush following withdrawal from the Kyoto Protocol in 2005, have actively tried to draw attention away from the UN process, although its successor, the Major Economies Forum on Energy and Climate Change (MEF), has since attempted to make closer linkages to the multi-lateral system.

Another important set of public sector transnational responses have emerged through networks of cities linking industrialized and developing country municipalities alike. Among them, the World Mayors Council on Climate Change (WMCCC) is notable in creating an alliance of advocates for city-based action linked to the multi-lateral response. Similarly, the C40 Climate Leadership Group represents a network of 58 of the world's megacities, containing one in 12 of the world's population, taking networked action to reduce greenhouse gas emissions. The Asian Cities Climate Change Resilience Network (ACCCRN) established by the Rockefeller Foundation is notable in its focus on adaptation rather than greenhouse gas-related issues in ten cities across Asia.

In the private sector, initial widespread resistance to the regulations of the international climate regime has given way to far greater understanding of the potential business opportunities from tackling climate change, even among some of the fossil fuel-based energy companies. Responses include business-leadership groupings such as those through the World Business Council of Sustainable Development (WBCSD), whose initiatives include working with 29 leading global companies from 14 industries to create a *Vision 2050*, which identifies a pathway to a 2050 world in which nine billion people can live well and within the planet's natural resource boundaries (WBCSD, 2010). Many multi-national corporations have moved from climate change as part of corporate social responsibility to developing voluntary regulation and target-setting to drive sustainable development concerns across the entire global operations (such as SABMiller – see Box 3.9). Authors on transnational governance have also drawn attention to hybrid models linking public, private and civil society sectors in public-private partnerships, including those such as REEEP renewable energy initiative that emerged following the Johannesburg summit of 2002 or those emerging from UNFCCC-related mechanisms such as the CDM or REDD+ (Bulkeley and Newell, 2010).

Box 3.9

Self-regulation and target setting: The case of SABMiller

Multi-national drinks manufacturer SABMiller includes *Energy and Carbon* as one of its ten priorities to 'make sustainable development part of everything we do'. It has set its own target to be 50 per cent more carbon-efficient than 2008 levels by 2020, defined in terms of emissions per litre of beer produced. SABMiller's carbon footprint results particularly from packaging manufacture, energy use in their breweries, and transportation. Actions to reduce greenhouse gas emissions are being undertaken through a combination of:

1 **Improving energy efficiency** in breweries, bottling plants and transportation.

2 **Investing in renewable energy sources** in its facilities.

3 **Switching to cleaner fossil fuels** where the switch to renewable is not easily achievable.

4 **Utilising carbon trading to reduce risk and create value**, both purchasing carbon credits to help meet carbon caps and pursuing appropriate opportunities for carbon credit creation.

5 **Moving to lower carbon packaging**, including retaining returnable bottle schemes where they exist.

6 **Reducing transport emissions**.

7 **Encouraging low emissions fridges** in distribution chains.

8 **Publicly reporting** greenhouse gas emissions.

Source: SABMiller, 2012.

Finally, there has also been an expansion of transnational partnerships between businesses and civil society, and a growth in internationally networked community-based initiatives on climate change. For example, community forestry has been engaged with payments for ecosystem services through both CDM projects (at least initially) and more recently with the development of REDD+ initiatives (Agrawal and Angelsen, 2012). The series of International Workshops on Community-Based Adaptation (CBA) initiated in 2005 (Ayers *et al.*, 2013) and the online knowledge platform *CBA-xchange* (http://community.eldis.org/cbax/) provide a strong example of the information-sharing function of transnational partnerships that can link communities, researchers and civil society with government and business.

Conclusion: A development diagnostic for the international climate change response

The UNFCCC continues to play a dominant role in global climate governance. This chapter has set out some of its main characteristics and mechanisms pertinent to a development context. It also provides a set of diagnostic questions around equity and effectiveness to allow the reader to assess the evolving future international regime from a development perspective. To date, the UNFCCC system has had only limited success in generating collective action consistent with climate stabilization, equity and a right to develop. As a consequence, the emphasis may be moving towards other local and transnational mechanisms and nationally determined and voluntary response that is framed by development priorities (see Chapters 4 and 5).

Given the evolving nature of the UNFCCC negotiations as they move towards an agreement for a comprehensive framework for 2020, a generic set of diagnostic questions is shown in Box 3.10 which can be applied to future proposals, agreements, policies and measures in order to assess them from a development perspective. This is grounded on the issues presented in this chapter which suggest that international climate change action must be based on negotiated principles of equity and the effectiveness of the response from the perspective of climate stabilization.

Box 3.10

A development diagnostic for the international climate change response

A development perspective on the collective international response to climate change can be considered in terms of both achieving climate protection and achieving development outcomes. Questions for assessing the response include:

Equity

● How is the total burden of emissions reductions allocated between and within countries?

● How does this burden reflect different responsibilities, capacities and the right to sustainable development?

● Is there effective international cooperation to assist mitigation and adaptation efforts in developing countries?

- How are countries and stakeholders engaged in the international decision-making process?

Effectiveness

- Is there a broad consensus on a stabilization target for the collective international response?

- Is the target consistent with the goal of preventing dangerous interference with the climate system?

- Can the mechanisms for international action deliver on these targets?

- Are systems in place to monitor and hold countries to account for progress?

- Can natural and human systems adapt to those impacts that cannot be prevented?

Summary

- The United Nations Framework Convention on Climate Change (UNFCCC) was established as part of the Rio Earth Summit of 1992 against the backdrop of growing international concern over global environmental problems.

- The overall aim of the UNFCCC is to avoid dangerous interference with the climate system. The UNFCCC initially reflected climate change largely as an environmental pollution-control problem and has gradually evolved to take a more development-oriented approach.

- The initial commitments of the Convention were strengthened by the Kyoto Protocol of 1997 but failure to agree legally binding actions at the Copenhagen conference of 2009 has since resulted in a voluntary approach and negotiations for a comprehensive agreement by 2015.

- Assessing the international response from a development perspective involves a combination of effectiveness in meeting the stabilization objectives of the Convention and equity with respect to sharing the costs of tackling the problem.

- A range of transnational, community-led and private sector responses have all emerged outside the auspices of the international climate change regime.

Discussion questions

- What are the advantages and disadvantages of the UN system in tackling climate change challenges?

- How might equity be re-evaluated and operationalized in a new global climate change agreement?

- What steps could be taken to improve procedural equity in the international climate change regime?

- How might we revise our international response to take account of dangerous levels of interference in the climate system?

- How might efforts outside the UN system be organized to meet the level of ambition required to prevent dangerous levels of climate change?

Further reading

Adger, W.N., Paavola, Y., Huq, S. and Mace, M.J. (eds) (2006) *Fairness in Adaptation to Climate Change*, Cambridge, MA: MIT Press.

Bulkeley, H. and Newell, P.J. (2010) *Governing Climate Change*, Abingdon: Routledge.

Ockwell, D.G. and Mallet, A. (eds) (2012) *Low-Carbon Technology Transfer: From Rhetoric to Reality*, Abingdon: Routledge.

Roberts, J.T. and Parks, B.C. (2007) *A Climate of Injustice: Global Inequality, North-South Politics and Climate Policy*, Cambridge, MA: MIT Press.

UNDP (2007) *Human Development Report 2007/2008 Fighting Climate Change: Human Solidarity in a Divided World*, New York: United Nations Development Programme.

Yamin, F. and Depledge, J. (2004) *The International Climate Change Regime: A Guide to Rules, Institutions and Procedures*, Cambridge: Cambridge University Press.

Websites

www.unfccc.int
Website of the UN Framework Convention on Climate Change, managed by the UNFCCC Secretariat. Contains all the agendas, negotiations decisions, submissions from country Parties, background papers on key issues, and tools and methodologies. Also hosts webcasts of the plenary meetings.

www.iisd.ca/process/climate_atm.htm
IISD Reporting Services produces summaries of the UNFCCC meetings and related
    events. Some reports include analysis, plus digital coverage with photographs and
    audio recordings.

www.climatenetwork.org/eco-newsletters
ECO is a daily insider's look at what is happening in the climate negotiations from the
    perspective of the international campaigning collective Climate Action Network
    (CAN).

www.climateactiontracker.org/
This 'Climate Action Tracker' is an independent science-based assessment which
    tracks the emission commitments and actions of countries. Maintained by a
    partnership of Ecofys, Climate Analytics and the Potsdam Institute for Climate
    Impact Research (PIK), the website provides an up-to-date assessment of
    individual national pledges to reduce their greenhouse gas emissions and their
    cumulative implications for global warming.

4 Mitigation responses and low-emissions development

In many cases mitigation and development can be mutually reinforcing.

- The development context for climate change mitigation
- Analytical tools to assess mitigation choices, policies and actions
- Overview of main mitigation options at the national level in main sectors relevant to development
- Concrete examples of key aspects of the mitigation challenge from a development perspective

Introduction

Whilst there is broad consensus internationally on the need to mitigate **climate change**, the question of what constitutes a fair sharing of the **mitigation** effort between countries and peoples – and in particular the role and contribution of poorer countries – remains contentious and divisive. Rich countries bear most of the historic responsibility for causing the problem and have far more resources to undertake mitigation actions. However, the magnitude of the challenge and the growing contribution of developing countries to global emissions mean that their actions alone will not suffice and developing countries will need to actively contribute to global mitigation efforts.

The urgency and magnitude of climate change mitigation contrasts with the slow pace of international and national responses. To date, pledged emissions reductions do not add up to what is required to reach agreed **stabilization** targets, and there is a further gap between pledges and actions. The **UNFCCC** process acknowledges the special needs and circumstances of developing countries, but there has been little impetus for integration of mitigation and development planning and policies at the national level. Consequently, and as a result of growing international awareness and concern, there has been increasing focus on autonomous national and voluntary approaches outside the framework of the UNFCCC.

In this chapter we consider mitigation responses in the context of national and local development rather than as a response to international rules setting. The first part of this chapter situates the mitigation challenge within the broader development context. It considers the relationship between development and emissions in historical and comparative terms, and identifies main trends relevant to mitigation efforts. The second section considers different analytical approaches to mitigation and their relevance to developing countries. The final section delves in to the strategic considerations and policy instruments and choices underpinning effective mitigation actions in developing countries.

1. The development context

In this section we consider historical patterns and cross-country comparisons in the relationship between development and **GHG** emissions, as well as the big development trends and their implications for efforts to control GHG emissions.

1.1 Development, economic growth and greenhouse gas emissions

The past two centuries have witnessed rapid and unprecedented progress in reducing poverty and improving human well-being. These improvements have been linked to economic growth, and concomitant increases in energy consumption and in **greenhouse gas** emissions. The close link between development and emissions has been evident from the beginning of the industrial age. Studies considering the relationship between GDP and emissions over multiple countries and time periods consistently find a close correlation between per capita GDP and emissions (Schmalensee *et al.*, 1998; Ravallion *et al.*, 2000; Heil and Selden, 2001). Figure 4.1 shows this trend for the years 1970 to 2005.

While global emissions have risen, the ratio of emissions to world GDP – the 'emission-intensity' of the global economy – has declined over the last hundred years. The amount of CO_2 emitted for every US$1,000 of GDP effectively halved from 1.27 tonnes in 1910 to 0.62 tonnes in 2006, linked to a decline in energy intensity of GDP. Much of this decline predates and therefore has nothing to do with **climate** policies. Three factors help explain this trend:

(a) gradual shifts in the composition of economic activity towards an increasing share of relatively low-emitting services sectors and a declining share of the more heavily polluting industrial and manufacturing sectors;
(b) increased energy-efficiencies linked to technological innovation and advances in production techniques;

(c) the development of cleaner energy sources such as natural gas, nuclear, hydropower, and more recently wind, solar, biofuels etc.

This 'automatic decarbonisation' (that is, reductions in emissions-intensity that have nothing to do with climate policy) has prompted some to question the urgency of mitigation, but such optimism is misguided. First, any improvements in energy (and carbon) intensity that were achieved were more than cancelled out by increases in the scale of economic activity over the same period. In technical terms, although there is clear evidence of 'relative' decoupling of emissions and economic activity, there is equally clear evidence indicating the lack of 'absolute' decoupling. Secondly, Figure 4.1. shows that the declining trend flattens out after the year 2000. In fact, there has been a trend towards *recarbonization* of the world economy over the past decade, driven in large part by the rapid economic rise of the big emerging economies, led by China and India.

In addition to humanity's growing consumption of fossil fuels, the steady depletion of terrestrial stocks of carbon stored in soils, trees and ecosystems has been a major source of greenhouse gas emissions,

Figure 4.1 *Trends in income, energy consumption and CO_2 emissions, 1970–2005.*

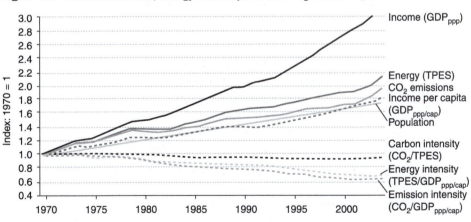

Source: IPCC, 2007.

Note: Relative global development of Gross Domestic Product measured in PPP (GDPppp), Total Primary Energy Supply (TPES), CO_2 emissions (from fossil fuel burning, gas flaring and cement manufacturing) and Population (Pop). In addition, in dotted lines, the figure shows Income per capita (GDPppp/Pop), Energy Intensity (TPES/GDPppp), Carbon Intensity of energy supply (CO2/TPES), and Emission Intensity of the economic production process (CO_2/GDPppp) for the period 1970–2004.

accounting for almost a third of anthropogenic emissions globally, and about 80 per cent of emissions for least developed countries (WRI climate analysis indicators tool, CAIT). The conversion of forests for agricultural use is the primary driver of land-based emissions; unsustainable logging practices in response to the growing demand for timber are also significant. Agricultural activities and the production of chemical inputs to agriculture (e.g. nitrogen fertilizers) are also major sources of emissions. Most of these emissions are of methane (CH_4) and nitrous oxide (N_2O) – which are far more potent greenhouse gases than CO_2.

There are large differences in carbon emissions between countries. Rich OECD member countries account for 15 per cent of the world's population and 45 per cent of global anthropogenic CO_2 emissions. With a slightly lower share (11 per cent) of world population, sub-Saharan Africa contributes only 2 per cent of global emissions. Low-income countries as a group account for a third of the world's population but for just 7 per cent of global emissions (UNDP, 2008; p. 57).

These inequalities are mirrored also in divergent energy consumption patterns. Currently, 1.6 billion people around the world lack access to electricity, some 85 per cent of them in rural areas. There are concerns that the 'brutal arithmetic' of climate change – whereby developing countries will be required to make deep cuts in their emissions in line with agreed stabilization targets – may reinforce existing inequities in energy access, thereby creating a situation of 'carbon apartheid', a situation that will only worsen as the **carbon space** continues to shrink (IEA, UNDP, UNIDO).

That economic growth and greenhouse gas emissions correlate does not signify a deterministic relationship, i.e. that emissions and development are irrevocably linked. In fact, data show a significant range in the carbon intensity of different countries at similar levels of economic development, suggesting real choice in terms of the emissions intensity of development pathways. This is clearer even if we look at a broader measure of human progress: the relationship between the **human development** index and greenhouse gas emissions is even less straightforward (see Figure in Box 8.1 in Chapter 8 of this book).

1.2 Population, energy and economic trends and implications for climate change mitigation

---

Box 4.1

Energy access for development: Current status and future challenges

The supply of reliable and affordable energy is a critical enabler of human and economic development: it provides for basic needs such as lighting, cooking, heating and transport, is an essential input to productive activities reliant on mechanical power, and underpins essential services such as the provision of clean water, sanitation and healthcare. As UN Secretary General Ban Ki Moon put it: 'Universal energy access is a key priority on the global development agenda. It is a foundation for all the Millennium Development Goals.' Below are some illustrative examples of how energy access contributes to the MDGs.

Eradicating extreme poverty and hunger (MDG1): Access to modern energy is strongly correlated with per capita income and can greatly increase the productive capacities of poor people. Modern energy can power water pumping, providing drinking water and increasing agricultural yields through the use of machinery and irrigation. Modern fuels reduce time spent collecting biomass fuels and increase time for income-generating activities. Refrigeration reduces food and crop waste.

Achieving universal primary education (MDG 2): In impoverished communities children commonly spend significant time gathering fuel-wood, fetching water and cooking. Improved cooking facilities free up time spent collecting wood, for education. Electricity is important for education because it facilitates communication, particularly through information technology, but also by the provision of such basic needs as lighting. Lighting also allows children to study at night.

Promoting gender equality and women's empowerment (MDG 3): Improved access to electricity and modern fuels reduces the physical burden associated with gathering and carrying wood and frees up valuable time, especially for women and girls, to gain an education or widen their employment opportunities. In addition, street lighting improves the safety of women and girls at night, allowing them to attend night schools and participate in community activities.

Improving people's health (MDGs 4, 5, 6): Improved access to modern energy services can contribute significantly to reducing child mortality, improving maternal health and reducing the incidence and effects of HIV/AIDS, malaria and other diseases. Reduced exposure to indoor air pollution from use of traditional biomass fuels for cooking and heating can significantly reduce mortality and morbidity, particularly among women and children. Improved access to energy allows households to boil water, thus reducing the incidence of waterborne diseases. Electricity and modern energy services support the functioning of health clinics and hospitals.

Ensuring environmental sustainability (MDG 7): Modern cooking fuels and more efficient cook-stoves can relieve pressures on the environment caused by the unsustainable use of biomass. The promotion of low carbon renewable energy is congruent with the protection of the environment locally and globally, whereas the unsustainable exploitation of fuel-wood causes local deforestation, soil degradation and erosion. Using cleaner energy also reduces greenhouse gas emissions and global warming.

Source: Adapted from UN-Energy, 2006.

Development will inevitably entail increases in GHG emissions for most poor countries. This increase will be driven primarily by growing populations, increased demands for energy to meet basic household needs and to fuel industrialization, infrastructure development and agricultural modernization.

Energy demand will soar in the coming decades, mostly in the developing world. World electricity demand is projected to double by 2030 (from a 2004 baseline) and on current trends (IEA, 2011). This steady growth in energy demand is explained by: (a) global economic growth: global GDP is set to double by 2030 and quadruple by 2050 from 2010 levels; (b) the importance of increasing access to energy and electricity to improving well-being (see Box 4.1); (c) the very low current rates of access to modern energy services in much of the developing world, e.g. only about a third of sub-Saharan Africans have electricity and 80 per cent rely on traditional use of biomass (Legros *et al.*, 2009).

Increasing the energy supply is not necessarily at odds with climate stabilization goals. Indeed the World Bank estimates that increases in global energy consumption of up to 57 per cent from current levels by 2050 are possible without compromising emissions reductions consistent with a 450ppm stabilization target. This would, however, call for profound changes to the energy mix – with the share of fossil fuels dropping from 80 per cent of energy supply today to 50–60 per cent (World Bank, 2010a).

More significantly, the available **carbon budget** can easily accommodate an expansion of basic modern energy services to all. Increasing access to electricity services and clean cooking fuels in South Asia and sub-Saharan Africa would add less than 2 per cent to global CO_2 emissions (World Bank, 2010a).

Whilst expanding access to modern energy services for basic needs is compatible with climate stabilization objectives, most of the increase in energy demand from developing countries in the next decade will be driven by the rise of a small handful of big emerging economies (e.g. China, India, Indonesia, Brazil, Mexico) and the growing needs of a rapidly expanding middle class associated with that rise. As a result, the global share of developing countries' emissions is set to rise, and soon to overtake OECD countries', underscoring the necessity of developing countries' participation in the global mitigation effort.

An important point to underline here is that extending access to basic energy services for the poor is a very different issue and challenge from satisfying the ever-increasing energy demands of a growing middle class. One is about poverty eradication, the other is about consumption choices.

Population growth has been an important factor in emissions increases, with global population roughly quadrupling over the past century. In the future, it is not so much the growth in population but the growth in consumption that will drive the growth in greenhouse gas emissions as most of the expected increases in population will be concentrated in poor regions characterized by consumption levels so low that they contribute little to emissions growth (Satterthwaite, 2009).

Rapid urbanization and the growth of cities is another important force shaping our climate future. The way cities are shaped and run has huge implications for greenhouse gas emissions and development. The world's cities now consume between 60 to 80 per cent of the world's energy and account for a roughly equal share of global CO_2 emissions (Kamal-chaoui and Robert, 2009). The next 20 years will see unprecedented urban growth – from three billion people to five billion, mostly in the developing world (*ibid*). Yet the concentration of population in urban localities also represents a critical opportunity for innovation and sustainability through efficient urban design, sustainable transport, green buildings etc. (see section 3.4 of this chapter).

2. Analytical approaches to climate change mitigation and low-emissions development

Since almost all human activities generate greenhouse gas emissions in varying measures, there are potentially limitless ways one could go about mitigating climate change. However, climate change mitigation competes with other immediate concerns – e.g. health and education, food and energy security, employment generation – for political attention and limited public resources. 'Where to start?', 'What mitigation options to prioritize?', and 'How far to go?' become central questions for public policy and decision-making. Here we will consider some analytical approaches and frameworks that assist in identifying and ranking relevant mitigation actions, and assessing trade-offs and synergies with development objectives.

There are many competing – and sometimes contradictory – perspectives on what constitutes effective mitigation action. For example, expanding nuclear power might be favoured by engineers on the basis of their considerable technical abatement potential, questioned by an economist on the basis that lesser cost options are available, and rejected by political leaders and advisers because of a combination of the high cost, public fear of nuclear safety, and resistance from vested interests such as fossil fuel lobbies.

Perspectives also differ according to institutional mandates, niches and imperatives. For example, the multi-lateral development banks have tended to focus on large-scale energy efficiency and renewable energy investment schemes in middle-income countries (MICs), which may have significant mitigation potential but which only indirectly address poverty reduction concerns. On the other hand, leading bilateral development agencies (e.g. DFID, Danida, Norad) and development NGOs tend to put much more emphasis on mitigation activities that directly contribute to poverty reduction in poorer countries.

While it is important to recognize and bear in mind these different perspectives, our emphasis here – as in the rest of the book – is on mitigation in the context of national development priorities. Broadly speaking, analytical approaches can be thought of as falling under three main categories, which are overviewed here in turn: (a) technological/engineering approaches; (b) economic/management approaches; (c) development first approaches. First, though, let us consider how to break the problem down at a general level.

2.1 The Kaya Identity and the basic maths of reducing energy-related emissions

A useful starting point for thinking about mitigation responses is to decompose emissions into its main determinants in order to identify where the broad policy levers for effective action may lie. The Kaya Identity – a simple equation that sets out the different factors that determine levels of carbon emissions from human sources – provides a helpful and commonly used framework for doing that in the context of energy-related emissions, which are growing rapidly and form a large proportion (70 per cent) of overall emissions (Raupach et al., 2007).

The identity relates total anthropogenic carbon emissions (C) to four factors: global population (P); per capita economic output expressed in terms of per capita world GDP (g); energy intensity of economic output (e), expressed as primary energy consumption per unit of output; and carbon intensity of energy (c), expressed as carbon emissions per unit of energy consumed:

$$C = P * (G/P) * (E/G) * (C/E) = P*g*e*c$$

where G = world GDP and E = total global primary energy consumption.

The Kaya Identity tells us that there are four ways to reduce energy-related emissions: by reducing population, producing/consuming less, increasing the efficiency of energy use (i.e. delivering more economic output at any given level of energy use), or reducing the reliance on carbon-rich fuels by switching to lower carbon energy sources (e.g. renewable energies).

Population management, while politically contentious, certainly has a place in the discourse on climate change mitigation, not least as improved access to family planning and reproductive health services can deliver significant social benefits – particularly for women – in addition to avoidance of future emissions. That said, population management measures cannot be expected to do much more than reduce the rate of population *growth*. In other words, they are unlikely to deliver absolute emissions reductions, at least not in the time frames that matter for climate action.

The second set of options involves reducing economic activity. Population growth alone means that in order just to keep per capita income levels constant, total GDP will need to continue growing.

Despite the efforts of sustainable consumption campaigns, decreases in per capita income are unlikely to be palatable to developing and richer country governments alike.

The last two terms in the Kaya Identity, energy efficiency and a shift to lower carbon energy sources, are linked to the emissions-intensity of GDP, which is usually related to fuel switching and technological upgrading. The main conclusion to emerge from a basic Kaya Identity analysis therefore is that CO_2 emissions reductions will have to come primarily from technological changes leading to greater energy efficiency and substitution of clean for dirty energy sources.

The Kaya Identity is widely used as a way to think about the drivers of emissions at the broadest level, including in the development of future emissions scenarios (such as the IPCC Special Report on Emissions Scenarios (SRES)).

2.2 Techno-centric approaches and the 'stabilization wedges'

While the Kaya Identity highlights the central role of technological change in mitigating climate change, it does not provide any practical guidance as to what technological options to pursue and to what degree. Starting with the estimated emissions reduction required by 2054 consistent with the 2-degree stabilization target, Pacala and Socolow (2004) break this total into 15 discrete action 'wedges'. These include improvements in energy efficiency, expansion of renewable energies, fuel switching to cleaner fuels (e.g. coal to natural gas), enhancement of natural sinks (e.g. adoption of conservation tillage in agriculture, reduced deforestation), nuclear energy, carbon capture and storage and increased use of biomass fuels.

This approach has been adapted by others, including the World Bank in its 2010 World Development Report (see Figure 4.2). They take us a step further than the Kaya Identity in that non-energy sectors are included in the analysis, and main technological options are identified and quantified. And they provide a useful illustration of how a full range of mitigation measures need to be used concurrently to address climate change.

The wedges approach provides a top-down, technologically driven analysis, which may help frame and guide the debate at the global

Figure 4.2 *Envisioning emissions reduction options through 'stabilization wedges'.*

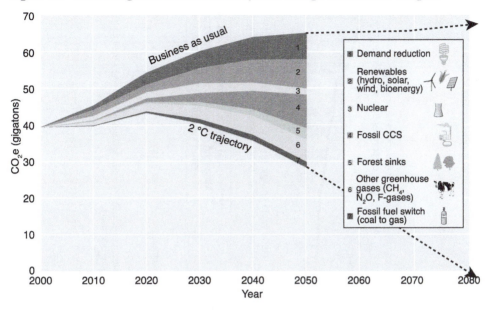

Source: World Bank, 2010a.

level, but provides little practical guide for developing countries about which actions should be prioritized at the country level.

2.3 Economic approaches and the Marginal Abatement Cost (MAC) Curves

Cost-effectiveness is a central consideration in prioritizing mitigation actions. The marginal abatement cost (MAC) curve provides a visual representation of the size of potential technical GHG abatement measures for different activities in order of cost. A global MAC curve is shown in Figure 4.3. Potential emissions savings (abatement) are on the x-axis and the cost (per metric ton of carbon dioxide) is represented on the y-axis. It thus allows us to see, *for any given price of carbon*, the range of mitigation interventions that would be incentivized economically (i.e. that would pay for themselves through cost-savings).

It also shows us that even in the absence of a carbon price there would be a range of mitigation actions that would make economic sense. Globally, these negative-cost measures, which are mainly linked to

Figure 4.3 *The global GHG marginal abatement cost curve.*

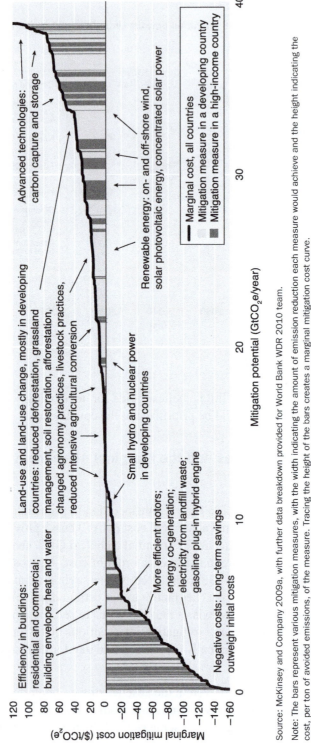

Source: McKinsey and Company 2009a, with further data breakdown provided for World Bank WDR 2010 team.

Note: The bars represent various mitigation measures, with the width indicating the amount of emission reduction each measure would achieve and the height indicating the cost, per ton of avoided emissions, of the measure. Tracing the height of the bars creates a marginal mitigation cost curve.

efficiency measures, could add up to more than ten $GtCO_2e$ of abatement per year (McKinsey and Co., 2009a).

The cost curve highlights the important potential contribution of land-based mitigation options, including reducing slash-and-burn agriculture conversion, degraded land restoration, pastureland and afforestation. These may be particularly relevant in poor, pre-industrial developing countries, where terrestrial sources constitute a large share of national emissions.

The use of MAC curves has become common for prioritizing mitigation actions in national and sectoral plans. National MAC curves vary greatly in both shape and content, depending on endowments of fossil and renewable energy resources and of forests, their economic structure and the sectoral pattern of emissions, which in turn are linked to income levels. Low Income Countries (LICs) use relatively little energy, and typically about half of their emissions tend to come from agriculture (or forests if relevant). By contrast, in fast growing, diversifying middle-income countries, a much higher proportion of emissions and abatement opportunities (in the range of 60–80 per cent) come from the energy, industry and transport sectors (WRI Climate Analysis Indicators Tool, CAIT).

The MAC analysis focuses only on technical feasibility and narrowly defined economic costs. Considerations of the (often considerable) transaction and implementation costs do not enter into the analysis, nor do the ancillary benefits associated with mitigation actions (known as 'co-benefits'). For example, energy efficiency measures may have associated benefits in terms of both energy security and improved health outcomes from reduced local pollution.

In addition, political economy factors – how trade-offs are negotiated and managed, who gains and loses out from proposed mitigation actions, and hence is likely to support or oppose these, and how power is exercised in political decision-making – will also play an important role in determining options. Even when local benefits far outweigh the costs of action, mitigation measures may be constrained by political and institutional factors. This explains why some actions that lie under the MAC curve – negative cost options – are not already being done. Overcoming these factors requires thorough analysis of the political context, followed by effective management of implementation as an inclusive political process.

2.4 'Development first' approaches and the integration of mitigation and development

The approaches described above regard the problem as largely technological and economic. They set out the realm of the possible but do not help in defining the sub-set of interventions that are particularly relevant and beneficial for developing countries and that should be prioritized. They have in fact contributed to a compartmentalized view of mitigation as a technical problem requiring technical solutions, overlaying but not directly interacting with broader societal and policy choices and concerns. This has led to the mitigation challenge being framed in terms of *additional* costs or burdens on societies, rather than as integral to development policy. The dominance of these approaches has been both a cause and a symptom of the institutional segmentation of mitigation concerns from the development process.

'Development first' approaches take a different starting point: mitigation is considered through the lens of development priorities rather than in terms of abatement options and potential. Such approaches, variously referred to as **low-emissions development** or **low-carbon development**, consider how domestic policy objectives – e.g. poverty reduction, energy access and security, improved urban transportation – can be achieved at minimum carbon-cost over time. In other words, they seek to identify lower-emissions pathways to achieving development objectives. These approaches are based upon a broader consideration of the costs and benefits of mitigation actions, taking into account social and environmental externalities, and institutional and governance realities. They thereby provide a practical framework for systematic integration, including identifying synergies, managing trade-offs and overcoming barriers to the implementation of solutions.

The rationale for low-emissions development (LED) as opposed to an abatement-driven mitigation policy is clear from a development standpoint: (a) scarce resources are concentrated on pressing development needs and additional climate finances are geared towards meeting these needs; (b) countries are explicitly recognized for their contribution to climate protection through sustainable development choices; (c) development decisions are informed by a fuller assessment of costs and benefits; and (d) development choices are as a result better adapted to an increasingly carbon-constrained world (Bradley and Baumert, 2005).

Even from a mitigation standpoint alone, 'development first' approaches have three major advantages over mitigation-led ones (Bradley and Baumert, 2005):

(a) stronger incentives exist to implement the necessary laws and policies, since mitigation actions would leverage existing policy priorities. This is a particularly important point in the context of mitigation actions as it is sustained effort over long timeframes (sometimes decades) which will make the difference, not a one-off policy decision;

(b) significantly larger flows of development finance can be mobilized in support of mitigation actions, rather than these being solely reliant on the smaller, more circumscribed and often difficult to access sources of climate finance;

(c) mitigation concerns can have a greater and more direct influence on development plans and policies, as they are integrated into development decision-making.

This approach supports the notion that development choices can make a major contribution to climate goals. Rather than isolated, incremental and often under-funded greenhouse gas abatement actions divorced from development choices, 'development first' approaches offer the prospect for shifting development trajectories in the direction of a lower emissions-intensive pathway over time, while maintaining a clear focus on development priorities along the way.

3. Policy options for mitigation in the context of low-emissions development

The previous section considered different ways to *think* about mitigation; here our focus is on what countries can *do* to mitigate climate change in the context of low-emissions development. Mitigation options are not limited to technological interventions, although these are certainly important, but relate to the broader policy agenda. This section provides an overview of strategic considerations and policy instruments for mitigation and low-emissions development in developing countries.

We start off by looking at the main 'instruments' of national mitigation policy and low-emissions development. We then consider how mitigation concerns interplay with development priorities in key sectors relevant to developing countries and poor people: energy, agriculture, forestry, transportation and urban development. Finally we look at how climate change mitigation can be integrated into national development planning and policies.

3.1 Policy instruments for climate change mitigation

Policy instruments are the tools that governments can use to achieve policy ends. For any given mitigation objective there will likely be a variety of policy instruments available to choose from. A useful way of categorizing policy instruments is to distinguish between four broad (and sometimes overlapping) categories: 'Using markets', 'Creating markets', 'Regulations', and 'Information and public engagement' (World Bank, 1997).

Using markets is about adjusting relative prices to reflect external costs. This is done through taxes, fees or by creating a situation of scarcity in a market via regulatory means (e.g. capping emissions). Carbon pricing is a powerful means of encouraging the private sector to develop and deploy least-cost, low-carbon technologies to achieve deep emission cuts.

While a uniform global price on carbon would be the most efficient solution, in practice this would require an unprecedented level of international cooperation. In the absence of a global carbon price, countries have to consider the impacts on competitiveness and risks of carbon leakage resulting from carbon price differentials between countries. Border carbon tax adjustments to correct for the under-pricing of carbon in countries of origin have been put forward as a means of addressing these concerns (Burniaux *et al.*, 2010).

In reality a uniform carbon price may be difficult to introduce even at the national level, particularly in developing countries. A practical alternative is to target the most heavy-emitting sectors (e.g. the power sector and the cement and steel industry) or products (e.g. petrol, Sport Utility Vehicles etc). China's introduction in 2007 of export taxes on 40 emissions-intensive products is one example of this. Targeted subsidies can be an important means to encourage the development and deployment of clean technologies and renewable energies. Conversely, reducing or eliminating subsidies to carbon-rich fossil-based energy use can help reduce emissions as well as government expenditure on fuel subsidies. This has been applied for example in Nigeria, Thailand, Senegal and Ghana.

Creating markets is about clearly delineating property rights so that markets can emerge where they may previously have been 'missing'. In developing countries, the actual creation of private property rights for land and other natural resources (e.g. forests, minerals) can be an important means to address key drivers of greenhouse gas emissions such as deforestation and slash-and-burn agriculture. Tradable emissions permits and rights and international offset mechanisms such as the Clean Development Mechanism (CDM) are further examples of market creation for mitigation.

Regulations such as bans, standards, quotas, and non-tradable permits have the advantage of providing a greater degree of certainty over final outcomes and have been successfully employed elsewhere, for example in reducing ozone-depleting substances. Regulatory approaches can be relatively costly as their effectiveness is premised on the existence of credible mechanisms for enforcement, including compliance monitoring and penalties. Market enthusiasts argue that imposing uniform requirements on activities with different abatement costs, and prescribing specific abatement technologies, can stifle cost-reducing innovation.

Information and public engagement – a major obstacle to effective action is the low awareness among policy makers and the public about options for and costs of emissions reduction, as well as the broader environmental and socio-economic co-benefits of low-emissions development. Specific measures include labelling, certification and stakeholder participation in the mitigation policy process. As noted further below, informing and educating farmers about best management practices can be a powerful means in itself of advancing agricultural mitigation.

A mix of these instruments can allow decision makers flexibility in finding ways to comply with government policy at the lowest possible cost. However, the possibility also arises of overlapping instruments unnecessarily adding rigidity, complexity and costs to compliance and in some cases undermining the achievement of policy goals. Finally, strong institutional support (coordination and capacity development) is needed to ensure overall coherence of the instrument mix and effectiveness of their implementation. For example, while environmental ministries are typically charged with mitigation policy, they rarely have the mandate to introduce taxes or subsidies. Close coordination with ministries of finance will therefore be important.

Actions should also take account of the dependence on or sensitivity of mitigation policies in particular sectors to decisions made in other sectors. For instance, realizing the mitigation potential associated with increased use of biomass resources will depend to a large extent on agricultural policies. Comparing environmental and social co-benefits and costs with the carbon benefit will highlight trade-offs and synergies, and help promote sustainable development.

3.2 Mitigating energy-related emissions

Tackling energy-related emissions is a central plank of any low-emissions development strategy, although the priorities, policies and measures will vary significantly depending on development context and national priorities.

The long lifespan of energy infrastructure – including power supply systems, and the factors determining energy demand such as buildings, factories, city design, and transportation systems – presents both a challenge and an opportunity for mitigation. Failure to bring mitigation concerns into energy policy at an early stage may create lock-in to inefficient and dirty energy infrastructure for decades to come and render future mitigation efforts more difficult and costly. Conversely, introducing efficient, low-carbon technologies into new energy stock can lay a long-term foundation for low-emissions development. There is therefore a strong premium on early action when it comes to mitigating energy-related emissions.

Mitigation actions will be considered alongside and balanced against energy policy priorities, which usually feature a mix of: (a) *supporting the growth of key economic sectors*, including manufacturing and other energy-hungry industries such as construction; (b) *expanding access to modern energy services*, e.g. electricity and modern fuels; (c) *energy security*, e.g. through energy conservation, diversifying energy sources, developing domestic energy resources.

The scope for improving energy efficiency in developing country economies tends to be large, both on the demand and supply sides in power, industry, buildings, and transport. These abatement options tend to be large and relatively cheap; in fact many of them are associated with negative marginal abatement costs, and significant co-benefits linked for example to improved health outcomes from reduced air pollution. Renewable and clean energy solutions also have significant mitigation potential, although some technological options

Plate 4.1 *Concentrated Solar Power plant in Sevilla, Spain.*

Credit: Koza1983 (own work by uploader) [CC-BY-3.0 (http://creativecommons.org/licenses/by/3.0)], via Wikimedia Commons.
Note: More and more governments are showing interest in investing in clean energy sources. With 3,000 hours of sun per year, concentrated solar power has the potential to make a major contribution to the energy mix in Morocco. In partnership with the World Bank, the European Commission and the Africa Development Bank the Moroccan government has undertaken to make Ouarzazate the home of what will be the largest concentrated solar power (CSP) plant in the world. At 500MW, the Ouarzazate CSP plant would produce 25 times the amount of power generated from the Sevilla CSP plant. With the first 120–160 MW plant Morocco would avoid 240,000 tons of CO_2 emissions a year – the same as removing 80,000 cars from the road annually whilst creating up to 11,000 jobs locally in CSP construction and maintenance

may be prohibitively costly (e.g. nuclear) and others have yet to be commercially proven (e.g. carbon capture and storage) (Lockwood and Cameron, 2011).

Most energy efficiency improvements also bring important other benefits in terms of energy security, industrial performance and competitiveness. The situation with renewable and clean energy technologies is more complicated though. On one end of the spectrum, bringing off-grid renewable energy solutions to rural areas will deliver great development benefits but yield small emissions reductions given the low levels of energy consumption by the poor. On the other end of the spectrum, investing in grid-connected renewable energy such as hydropower, nuclear power or carbon capture and storage can result in large emissions reductions but at a relatively high cost and may not serve the poor due to their higher cost.

Pricing reform is crucial to levelling the playing field for energy efficient and clean energy technologies. Where it entails phasing out existing fossil fuel subsidies, distributional impacts need to be carefully considered, as the poorest tend to be hit hardest by higher fuel prices. The experience of Ghana (see Box 4.2) shows how supplementary measures, such as targeted cash transfers to low-income households, can be crucial to enable the poor to cope with rising energy bills and build broad-based support for reform. In other cases, levelling the playing field entails introducing new subsidies to internalize external mitigation benefits. For example, feed-in tariffs – which guarantee a sufficiently high purchase price for renewable energy for them to compete with fossil energy – have been very effective in supporting the scaling up of renewable energy.

In addition to energy pricing reform, supporting regulations such as efficiency standards and codes – combined with institutional reforms and consumer education – will be important to overcoming informational and other barriers to the rollout of energy-efficient and renewable energy solutions. Brazil's experience provides a compelling example of how energy efficiency standards for buildings, electrical appliances and vehicles can dramatically curtail emissions at low cost.

Expanding access to modern energy services will be a cornerstone of energy policy in most developing countries. Impacts on emissions can

Box 4.2

Fuel subsidy reform in Ghana

In 2004, Ghana was spending around 3 per cent of GDP – well in excess of health spending – on subsidizing fuels and maintaining the state-owned refinery. Two earlier attempts to phase out fuel subsidies failed in the face of popular opposition. In 2005 the government embarked on a new round of reform. However, this time it built a political strategy. A poverty and social impact analysis (PSIA) was undertaken to reveal the extent to which wealthier Ghanaians benefitted from the existing subsidy structure, and the results used in debates on the issue. A number of steps were taken to compensate poorer households, including eliminating school fees, capping bus fares, and more resources for rural health care and electrification. A cross-subsidy for kerosene was maintained. This round of reform was successful and there were no street protests.

Source: Laan *et al.*, 2010: 11–14.

vary, but where emissions increase, the increases tend to be negligible while the development benefits are considerable (Lockwood and Cameron, 2011). Two main routes to expanding energy access are: (i) grid extension; (ii) off-grid, small-scale, decentralized energy solutions. The latter tend to be more cost effective. Feeding a mini-grid with a combination of different energy sources addresses the limitations of individual technologies and can provide a promising approach to rural electrification (Legros *et al.*, 2009).

3.3 Mitigating land-based emissions

Land-based mitigation options are amongst the most promising in terms of mitigation potential and development co-benefits but also amongst the most challenging to implement. They represent the only means of avoiding substantial increases in non-CO_2 greenhouse gas emissions in the coming decades (Smith *et al.*, 2007; Norse, 2012; Nabuurs *et al.*, 2007; Rose *et al.*, 2011) and have the potential to be a steady and important part of the portfolio of mitigation strategies in developing countries, particularly agricultural and rainforest nations.

They have gained momentum under the agendas of 'Reduced Emissions from Deforestation and forest Degradation' (REDD+, see Chapter 6) and 'Climate Smart Agriculture' (Angelsen *et al.* 2012; FAO, 2010).

Many land-based mitigation options can deliver large social and environmental co-benefits (see Box 4.3). Measures to promote tree and forest planting, improve forest management, cropland management, agroforestry, grazing land management, and re-vegetation can enhance ecosystem services for local livelihoods and industries while sequestering carbon (Nabuurs *et al.*, 2007; Smith *et al.*, 2007). And low-emissions agriculture can benefit the poor by increasing their food security, boosting their incomes and enhancing the productivity of their natural assets (e.g. through improved nutrient and water use efficiency) (Braun *et al.*, 2005). Moreover, there are strong potential synergies between carbon sequestration and **adaptation** measures, such as afforestation of vulnerable areas, watersheds, and rehabilitation of degraded lands.

Any land-based mitigation policy or measure must consider synergies and resolve trade-offs and conflicts between different land uses in the context of national development with particular regard to the needs of the poor. This is particularly important where the mitigation measure involves a major change in land use as opposed to protecting and

Box 4.3

Sustainable agriculture and forestry project in Nhambita, Mozambique

The Nhambita Community project in Mozambique provides an example of the potential for sizeable poverty reduction benefits from mitigation activities. Following years of devastation caused by civil war, the Nhambita project aimed to rehabilitate the Gorongosa National Park environment, reduce CO_2 emissions, and reduce poverty by incentivizing local people to adopt sustainable agricultural and forestry practices. Activities included:

- Adoption of agro-forestry practices helped improve the fertility of the soil, thereby increasing crop yields, reducing the need for nitrogen fertilizers, and enhancing the natural carbon absorption of the soil.

- Afforestation and planting of other crops led to enhanced sequestration of carbon. Local people are paid to plant trees and crops appropriate to the local habitat and maintain the land. Concurrently, the sustainable harvest of crops and trees provides a supply of fuel wood and other forest products.

The Nhambita community undertakes the sustainable practices under contract with Envirotrade, an organization that brokers the consequent carbon reductions. The carbon credits are independently verified, then purchased by organizations such as the Carbon Neutral Company on behalf of people who want to offset their emissions on a voluntary basis. The sustainable practices adopted by people in Nhambita are estimated to save 90 t CO_2 per hectare. However, there has been more limited success in accrediting small-scale sustainable agriculture and forestry initiatives as projects under the Clean Development Mechanism (CDM) because of the high transaction costs.

Source: Girling, 2005, cited in Stern, 2007.

maintaining existing carbon stores. In these cases policies and their impacts will need to be considered in the context of other, possibly competing needs for and uses of land: for food production, as living space, for maintenance of biodiversity, for recreation and to fulfil aesthetic and spiritual demands (Millennium Ecosystem Assessment, 2005).

Land-based mitigation covers both emissions abatement and removals (through sink enhancement) of greenhouse gases resulting from land use, land-use change and forestry activities. It also covers multiple greenhouse gases, whereas in the energy sector CO_2 is the main concern. Land-based mitigation options pertain mainly to the forestry and agricultural sectors, and fall largely within the following categories:

● Emissions abatement measures:
  (a) *Reducing emissions from deforestation and degradation* (REDD) – this is particularly important in the big rainforest nations (e.g. Brazil, Indonesia, Democratic Republic of Congo). This is the forest mitigation option with the largest and most immediate carbon stock impact in the short term per hectare and per year globally (Nabuurs *et al.*, 2007).
  (b) *Reducing emissions linked to agricultural production* – e.g. improving the energy efficiency of nitrogen fertilizer production, switching to 'greener' fertilizers, improving livestock diets to reduce enteric fermentation (Norse, 2012).
  (c) *Displacing carbon-intensive fuels and products* – such as cement and construction materials, by substituting agriculture- and forest-derived biomass alternatives, thereby increasing off-site carbon stocks in wood products and reducing the carbon intensity of fuels used in other sectors. According to the IEA, modern biomass as fuel for power, heat, and transport has the highest mitigation potential of all renewable sources (IEA, 2008).
● Sequestration and sink enhancement measures:
  (d) *Forest conservation* – i.e. protection of existing carbon stocks in *natural* forests, especially primary forests, including those that face no immediate threat from deforestation and degradation but could in future be subject to land-use pressures. This is particularly relevant to countries and areas with high forest cover and low rates of deforestation;
  (e) *Sustainable forest management*, i.e. actions that safeguard and/or enhance existing carbon stocks in *working* forests, e.g. through sustainable harvesting techniques, payments for ecosystem service schemes etc.
  (f) *Enhancement of carbon stocks*, e.g. by increasing the forest area through afforestation and/or reforestation actions and through landscape-scale restoration of forest ecosystems.
  (g) *Increasing carbon sequestration on agricultural lands*, e.g. by limiting the loss of soil organic matter (SOC).

Agricultural mitigation can be among the cheapest options for mitigation, while improving food security by increasing food supply and farmers' incomes. Nitrogen fertilizers account for the largest share of energy inputs to agriculture and hence should be a major target for

Plate 4.2 *Terraced farms in China and Uganda: (a) China Loess Plateau; (b) Terraced farms in Uganda.*

(a)

(b)

Credit: (a) Alessandra Meniconzi; (b) WRI / James Anderson.
Note: Terraced farming is one example of climate-smart agricultural practices that can offer significant benefits for agriculture, climate protection and food security. Terraces play an important role in controlling soil erosion, maintaining soil fertility and boosting yields. Both China and Uganda offer success stories of terracing leading to immediate increases in yields and incomes through improved soil cultivation and irrigation systems.

agricultural mitigation. Although they are often subsidized to promote food security and to boost farmers' incomes, their overuse results in unnecessary emissions, lower farmers' incomes, exhausted soils and depressed yields over time (Norse, 2012).

A range of possibilities exist to **substitute low- for high-carbon inputs** to agriculture, e.g. replacing chemical fertilizers with organic fertilizers, green manures and anaerobic digester liquor fertilizers, or replacing synthetic pesticides with green alternatives (Conway, 1997). Several of these approaches are feasible, cost-effective, and justified on purely environmental grounds (e.g. to protect water quality). A large-scale switch to organic farming as a solution has been advocated by some but is controversial, and moreover much can be achieved in the absence of such a shift, e.g. by adopting best management techniques including integrated pest management (Conway, 1997).

In terms of carbon sequestration, the rehabilitation of degraded grasslands, conversion of degraded croplands to forests and agroforestry systems can build significant carbon storage capacities. Estimates made for the IPCC Fourth Assessment Report indicate that **improved management of croplands and grazing lands** provides the largest and most cost-effective measure for GHG mitigation. Over 95 per cent of this technical potential comes from sequestration through the build-up of soil organic matter (SOC). Improvements in fertilizer-use efficiency, tillage practices, and grazing-land management can contribute significantly towards that end, as can the restoration of degraded lands and the re-incorporation of crop residues (Smith *et al.*, 2007a; Lu *et al.*, 2009; Lal, 2011).

The forestry and agricultural sectors can make important contributions to displacing carbon-rich fuels in other sectors such as power generation, residential heating and transportation, through substituting biomass for fossil fuels. According to the IEA, modern biomass as fuel for power, heat, and transport has the highest mitigation potential of all renewable sources (IEA, 2008).

Many of the technical options for mitigating land-based emissions are tried and tested, and involve management changes and evolutionary changes in technology which can be introduced at low or negative economic cost. Introduction of **best management practices** in the agriculture and forestry sectors is often an effective means of reducing greenhouse gas emissions, enhancing soil carbon sequestration, whilst at the same time delivering significant benefits in terms of rural incomes, air and water quality management (Norse, 2012).

Despite their relatively low costs and significant co-benefits, and the large mitigation potential, land-based mitigation opportunities are far from being realized at present. This is partly due to the greater complexity, risks and uncertainties involved in land-based compared to energy mitigation. Managing land-based emissions in developing countries can be particularly challenging because of the following characteristics of land use, land-use change and forestry activities that set them apart from activities in other sectors:

- *The direct and heavy dependence of many of the world's poor on forests and agriculture* – as a source of livelihood, income and well-being. There are over a billion forest-dependent people globally and 70 per cent of the world's poor live in rural areas and depend on agricultural activities for subsistence as well as earnings (IFAD, 2011);

- *Monitoring and measurement is more complex, uncertain and costly* – there remains a significant degree of uncertainty regarding terrestrial sources of emissions and the abatement potential and the effects of abatement measures. Also, it may be very difficult to disentangle the impact of policies by governments or other institutions from the effects of other human and natural factors that drive changes in carbon stocks;
- *Leakage can occur and be significant* – for example when measures to avoid deforestation in one area simply lead to conversion of forests, wetlands or grasslands elsewhere. Leakage further contributes to difficulties of accounting and measurement;
- *Reversibility and non-permanence* – actions to enhance carbon sinks can be reversed through human activities, disturbances, or environmental change, including climate change (Watson *et al.*, 2000). The sudden Amazonia dieback of 2005 is a case in point;
- *Many factors drive land use and land-use change* – including the price signals from international markets, which can drown out any signals from domestic policy. For example, the devaluation of the Brazilian real combined with climbing international prices for soy in the early 2000s triggered a massive acceleration in the conversion of forests for soy plantations. Likewise, biofuel mandates in Europe and the US led to large-scale conversion of land to grow biofuel crops, and the rise in food prices since 2007–08 has driven increased foreign land acquisitions for agricultural production across Africa (Oxfam, 2012).

3.4 Mitigation in cities

Urbanization in the developing world represents both a risk and an opportunity for climate change mitigation. Although cities are home to about 50 per cent of the world's population, their residents are responsible for emitting more than 80 per cent of its greenhouse gases (World Bank, 2010c). Although sprawling and low-density cities can create long-term lock-in to wasteful patterns of fossil energy use, the concentration of people and resources in cities creates possibilities for economies of scale in deploying more efficient infrastructure and low carbon energy solutions. The correlation between high rates of urbanization and high per capita emissions reflects the fact that urban dwellers tend to be wealthier and to consume more than rural populations. Proactive strategies to manage urban development, therefore, can significantly limit the growth of emissions from increasing urbanization and consumption throughout the developing world.

Energy use is the main source of greenhouse gas emissions in cities. Cities use energy primarily to light, heat and cool buildings, to power industry and appliances, and to transport people. Because cities consume 75 per cent of the world's energy, policies to scale up renewables could significantly reduce cities' emissions (World Bank 2010c).Three key factors affect a city's energy use:

(a) *the efficiency of technologies and materials* – energy efficiency, particularly of buildings, vehicles, industry and electronic appliances, has a significant impact on the energy demand of a city. The energy efficiency of materials and technologies is especially relevant to developing world cities, where building codes are often lax or not enforced and automobile fleets are often decades old;
(b) *urban form* – dense, mixed-use urban development is in general more energy efficient than sprawling, single-use development. In compact urban areas, apartment dwellers save energy by sharing heating, cooling, electricity, and appliances with other tenants in their buildings;
(c) *individual behaviors and choices* – when living in large suburban houses and commuting long distances to work by car is the norm a city will be responsible for many more emissions than will be the case in densely populated cities served by good and affordable public transportation systems.

In addition to these 'direct' emissions caused by activities within the city's limits, city dwellers are responsible for the harder to calculate 'indirect' emissions from goods and services that are produced outside the city's limits.

Developing countries are rapidly building housing, transit, and service infrastructure to accommodate about 65 million additional urban residents each year worldwide (between 2005 and 2010). The volume of urban construction over the next 40 years could equal that which has occurred throughout history to date (National Intelligence Council, 2012). Swift policy action is required to avoid **lock-in effects** of unsustainable development and direct urban growth on a low carbon pathway.

Catalyzing low carbon urban development will require coordinated and proactive land-use and transport planning to manage growth and limit emissions. Maximizing efficiency through dense, transport-oriented development requires that city authorities implement a range of policies from often separated agencies. The relevant policy levers include implementing land-use zoning and property tax structures that

---

Box 4.4

Curitiba's successful experiment with integrated urban transport and planning

The introduction and steady implementation of an integrated urban planning programme in Curitiba from 1965 onwards allowed the city to grow eight-fold from 1950 to 1990, while maintaining 75 per cent of commute travel by bus – a much higher public transport modal share than in other big Brazilian cities – as well as little congestion. As a result, Curitiba uses 25 per cent less fuel than cities of similar population and socio-economic characteristics. Two characteristics of the programmes seem to have contributed particularly to its success: (i) integration of infrastructure and land-use planning; and (ii) the consistency with which successive municipal administrations have implemented the plan over nearly three decades.

Source: Najam *et al.*, 2007; Rabinovitch and Leitman, 1993.

---

incentivize concentrated, mixed-use development; building codes that maximize efficiency; and public transportation development to increase access to affordable, convenient transit. The example of the city of Curitiba in Brazil is illustrative in that regard (see Box 4.4).

Coordination of complementary policies is critical and often particularly challenging for cities in the developing world (OECD, 2010). For example, a new bus rapid transit (BRT) system is unlikely to succeed in reducing emissions from private car trips if urban development is not concentrated around BRT stops and transfer nodes.

Market-based measures are also emerging as key policy options for reducing a city's emissions. Several cities, mostly in the developed world, have implemented congestion charges to internalize the social and environmental costs associated with traffic congestion in central cities. Congestion charges in Singapore and London, for instance, charge automobiles for entering the city's centre. Cities are also turning to emission-trading schemes as ways of reducing emissions.

3.5 Mitigation and national development planning

In most developing countries, mitigation actions are still largely driven by the UNFCCC (e.g. **Nationally Appropriate Mitigation Actions**, or **NAMAs**) and related international climate financing mechanisms, rather than by national development priorities and a long-term vision for low-emissions development. The UNFCCC process plays an important role in laying the groundwork for effective future mitigation actions, by helping put into place internationally consistent standards and frameworks for accounting, measurement, and reporting of greenhouse gas emissions – and to a lesser degree of mitigation actions – at the national level.

Quantifying emissions, identifying their sources and their underlying driving forces is an essential component of greenhouse gas management and accountability mechanisms. Thus developing national greenhouse gas inventories and baselines is a necessary first step in the planning process. It is only by disaggregating total greenhouse gas emissions into their main sources that priorities can be set and mitigation options selected.

Effective mitigation requires credible commitments and sustained effort over a long timeframe. An overarching, strategic vision or goal is therefore crucial to guide development and mitigation policy decisions over the long term, galvanize consistent private responses, and move beyond piecemeal mitigation responses disconnected from the broader development process. Ethiopia's vision of a Climate Resilient Green Economy (see Box 4.5), which includes the goal of achieving carbon neutrality for the economy as a whole by 2030,

Box 4.5

Ethiopia's Climate Resilient Green Economy strategy

The Ethiopian Government launched its Climate Resilient Green Economy initiative (CRGE) in 2011. The CRGE strategy builds on the Growth and Transformation Plan (GTP), the Ethiopian government's ambitious development plan which sets the aspiration for Ethiopia to reach middle-income levels by 2025. It is based on three complementary objectives: fostering economic development and growth; ensuring mitigation of greenhouse gases (GHGs); and supporting adaptation to climate change.

The CRGE strategy indicates clearly that green growth is crucial to achieve the ambitious targets in the GTP in a sustainable way. It describes a new model of development that integrates key aspects of economic performance, such as poverty reduction, job creation, and social inclusion, with those of environmental performance, such as mitigation of climate change and biodiversity loss as well as ensuring access to clean water and energy.

Source: Federal Democratic Republic of Ethiopia. The strategy document can be downloaded here: http://www.epa.gov.et/Download/Climate/Ethiopia's%20Climate-Resilient%20Green%20economy%20strategy.pdf

offers a good example of a country that has set a clear and ambitious long-term vision to guide and integrate its development and mitigation choices.

The success of mitigation actions is dependent on and sensitive to development decisions taken across a range of socio-economic sectors. The IPCC Third Assessment Report concluded that the choice of development paths can lead to very different GHG emissions futures, and that charting a lower emissions development path will require major policy changes in areas other than climate change (Najam *et al.*, 2007). This points to the need for integrating mitigation concerns into the mainstream of development policy and planning. Such **mainstreaming** is particularly important because, although climate change mitigation is usually seen as the province of environmental ministries, the mix of policy instruments needed typically extends well beyond the remit of environmental authorities.

Against the backdrop of scarce public resources and persistent hunger and poverty, key criteria for prioritizing actions towards achieving low-emissions development in developing countries include: (a) focusing on the largest sources of GHGs with the lowest unit mitigation costs; (b) maximizing the social, economic and environmental benefits; (c) prioritizing to those measures that are progressive and pro-poor in their impact.

Effective mainstreaming of mitigation concerns in development decisions will require supporting institutional reform and building of mitigative capacities. Because an array of decisions are necessary to influence emissions and development paths, coordination between policies in several sectors and at various scales becomes a fundamental challenge. There will often be need for example for social policy to complement mitigation policies in order to address, neutralize or mitigate the adverse social consequences that may result from some mitigation measures (e.g. full-cost pricing for carbon-intensive goods). Technical capacities will not only need to be developed in the design of mitigation policies but maintained throughout to ensure continuous monitoring, learning and adjustment, which can be particularly important in the case of more experimental measures.

Finally, national planners and policy makers should be mindful of the limits of planning and the role of other actors, and should take deliberate measures to create space and enabling conditions for others to participate in a decentralized and spontaneous way in the national mitigation effort.

Conclusion

In a warming and carbon-constrained world, mitigation can no longer be seen in isolation from development choices. Not only is mitigation compatible with development, in many cases the two can be mutually reinforcing. However, to date mitigation actions in the developing world have been largely driven by the UNFCCC and associated mechanisms (e.g. Global Environment Facility projects, the Climate Investment Funds) rather than autonomous initiative, and they have been piecemeal and incremental as a result.

For developing countries to contribute more meaningfully to climate stabilization and bring mitigation into the mainstream of development, the mitigation actions they undertake must speak more directly to their pressing development needs – e.g. of food and energy security, access to energy, mobility, poverty reduction. In other words, they must be guided by a long-term strategy of low-emissions development. Their actions will also need to be encouraged and rewarded through the international system. Resource transfers will be needed to help developing countries build relevant capacities and leap-frog high-polluting phases to clean development technologies and low-emissions pathways (the theme of financing is further explored in Chapter 6).

There is no blueprint for low-emissions development. Technological change will be necessary but there are no straightforward technology fixes: the inescapable fact is that effective mitigation will require deep societal and economy-wide shifts and the full gamut of available options will need to be investigated and many solutions implemented concurrently. Continuous experimentation and learning-by-doing will therefore be a crucial element of effective mitigation strategies.

Summary

- Developing countries are responsible for a small part of the overall greenhouse gas stock, yet are likely to generate the bulk of future emissions increases. They are also the most vulnerable to climate impacts and thus have the greatest stakes in climate stabilization.

- The strong historical correlation between development and emissions does not imply a deterministic relationship. Evidence indicates the possibility of lower emissions development pathways.

- The rapid growth of the middle class in big emerging economies is the key driver of developing countries' emissions increases, not poverty reduction. Improving the lives of the poor, e.g. through expanding access to energy, will contribute little to global emissions increases.

- Climate policy alone will not deliver effective mitigation actions. Framing mitigation as part and parcel of sustainable development may lead to more effective mitigation actions in developing countries.

- There is a strong premium on early action to avoid lock-in to inefficient emissions-intensive structures that can render future mitigation efforts more difficult and costly for decades to come. This is particularly true with big 'hardware' decisions relating to energy, infrastructure and urban development.

- There are good reasons to undertake mitigation actions other than climate stabilization. These ancillary social and environmental benefits – known as co-benefits – can be significant and mitigation actions can sometimes be justified on the basis of these benefits alone.

- Success of mitigation instruments/actions in reducing emissions will depend on tackling profound market and governance failures. Policy instruments will have to strengthen the institutional alignment of economic actors and the public interest.

- There's no blueprint for low-emissions development. Mitigation actions need to be country-led and tailored to specific national circumstances.

Discussion questions

- Should mitigation be a priority, given the more immediate concerns of reducing poverty and hunger and improving health and education for today's generation?

- Does the need to curb global greenhouse gas emissions preclude poor countries from growing? How can poor countries' right to develop be squared with the need to curb global greenhouse gas emissions?

- Fairness dictates that countries undertake differentiated mitigation responses. What are the ways in which mitigation priorities may differ between rich and poor countries?

- How can mitigation options be identified and prioritized in a developing country context?

- Why can market forces alone not deliver the degree of mitigation action needed to avert climate disaster?

Further reading

Bradley, R., and Baumert, K. A. (2005) *Growing in the Greenhouse: Protecting the Climate by Putting Development First*, Washington DC: World Resources Institute.

Hansen, J., Sato, M., Kharecha, P., Beerling, D., Berner, R., Masson-Delmotte, V., Pagani, M., *et al.* (2008) 'Target Atmospheric CO2: Where Should Humanity Aim?', *The Open Atmospheric Science Journal*, 2.1: 217–231.

IEA, UNDP, and UNIDO (2010) *Energy poverty: How to make modern energy access universal?* OECD/ IEA, Paris: International Energy Agency.

Najam, A. *et al.* (2007) 'Sustainable Development and Mitigation' in Metz, B., Davidson, O.R., Bosch, P.R., Dave, R., and L.A. Meyer (eds) *Climate Change 2007: Mitigation. Contribution of Working Group III to the Fourth Assessment Report of the Intergovernmental Panel on Climate Change*. Cambridge, UK and New York, USA: Cambridge University Press.

Stern, N. (2007) *Stern Review: The Economics of Climate Change*, Cambridge: Cambridge University Press.

Urban, F and Nordensvard, J. (eds) (2013) *Low Carbon Development: Key Issues*, London: Earthscan.

World Bank (2010) *World Development Report 2010: Development and Climate Change*, Washington DC: World Bank Publications. Chapter 4.

Websites

http://cait.wri.org
The World Resources Institute's **Climate Analysis Indicators Tool (CAIT)** provides comprehensive and comparable databases of greenhouse gas inventories and other climate-relevant data, analysis tools, and dynamic maps that can be used to analyze a wide range of climate-related data questions. CAIT's primary purpose is to provide high-quality information, accessible through easy-to-use yet powerful and transparent analysis tools, to support the many dimensions of climate change policy.

www.iea.org

The **International Energy Agency** website provides access to global and national energy statistics and assessments and is a great resource for interactive data-visualizations.

www.climatesmartagriculture.org

This website of the **climate-smart agriculture** approach provides a useful resource for essential information on harnessing mitigation opportunities in agriculture, forestry and fisheries sectors. It also offers a space for researchers and practitioners to share information, experiences and views on promoting climate-smart agriculture.

http://unfccc.int/national_reports/items/1408.php

National Communications – These provide information on greenhouse gas emissions and steps taken to comply with UNFCCC commitments. All parties to the UNFCCC are required to submit 'national communications' to the Conference of the Parties, through the secretariat, detailing the following: (a) A national inventory of anthropogenic emissions by sources and removals by sinks of all greenhouse gases not controlled by the Montreal Protocol; (b) a general description of steps taken or envisaged by the Party to implement the Convention; (c) any other information the Party considers relevant to the achievement of the objective of the Convention and suitable for inclusion in its communication, including, if feasible, material relevant for calculations of global emission trends.

5 # Adaptation responses and building climate resilient development

Adaptation needs to address the multiple climate-related hazards.

Introduction

This chapter introduces the concept and practice of **adaptation** to **climate change** in the context of development. Until the late 1990s, the focus of international debate and policy responses to climate change was on **mitigation**, and adapting to the impacts of climate change was almost seen as an admission of failure. The subsequent emergence of adaptation issues has been driven largely by developing countries, concerned that their geographical exposure, natural resource-dependent livelihoods and economies, and weak capacities to respond made them particularly vulnerable to **climate**-related shocks and stresses. The increased attention and resources now attached to adaptation is a result of developing countries' growing voice in the international climate arena, and adaptation has become a key demand of developing countries in the climate negotiations.

Nevertheless, adaptation has generally proved more difficult to undertake than mitigation. In part this is because its objectives cannot be reduced to a common measure (as **greenhouse gas** emissions can) and in part because of the difficulty of distinguishing between some forms of adaptation and standard development interventions (see Chapter 2). Multiple factors contribute to building capacity for adaptation (**adaptive capacity**), including wider development concerns such as education, health and governance. While adaptation can build on the historic experience of adapting to variations in the climate, it will also entail dealing with new conditions that have not previously been experienced, such as sea level rise and glacial retreat, and increases in and changing locations of extreme weather such as heat-waves and cyclones. A basic requirement for adapting to climate change is therefore to understand how different sectors, activities, infrastructure and ecosystems respond to changes in the climate, and what actions can be taken to reduce negative impacts.

This chapter starts by unpacking the concept of adaptation and its context in developing countries, before presenting a typology of four different approaches to analysing and implementing adaptation.

1. Understanding adaptation: Definition and anatomy

Historically, the theoretical concept of adaptation is associated with Darwinian evolutionary theory and seen as an integral part of the process of natural selection. The modern theoretical foundations of climate change adaptation draw in particular from work in the fields of **vulnerability**, livelihoods, **disaster risk reduction (DRR)** and **resilience**. Its use in climate change lexicon stems from the **UNFCCC**, although the most widely adopted definition comes from the IPCC Third Assessment Report (IPCC, 2001a) as an 'adjustment in natural or human systems in response to actual or expected climatic stimuli or their effects, which moderates harm or exploits beneficial opportunities'.

This definition highlights the potential opportunities as well as more commonly held concerns about risk and impact mitigation linked to a changing climate. Adaptation in the development context is usually focused on protecting and enhancing people's well-being and livelihoods in the face of a changing climate as part of ongoing processes of economic growth and poverty reduction.

Barry Smit and colleagues (2000) suggest an anatomy of adaptation to climate change that provides a useful checklist for considering adaptation responses across multiple contexts. They suggest four key guiding questions: (a) Adaptation to what?; (b) Who/what adapts?; (c) How does adaptation occur?; and (d) How good is the adaptation?

1.1 Adaptation to what?

First one must ask what aspect of climate change is the focus of the proposed adaptation measure. The term 'climate change' is often used generically without clarifying if this refers to specific parameters such as rainfall, temperature, or wind, or to specific impacts such as glacial melting, extreme events, sea level rise, crop failure or flooding. For landlocked African countries such as Ethiopia, Niger, Malawi, Lesotho or Zambia, sea-level rise may not be a major concern, for example, but as agriculture is predominantly rain-fed, variations in rainfall will be an important concern.

This initial question also helps to clarify if the adaptation is to long-term average changes in climate conditions (e.g. temperature and sea level rises), changes in climate variability such as seasonal shifts, or to extreme events that may occur with greater frequency due to global warming (see Figure 1.3 in Chapter 1). Increasingly, climate science suggests the need to consider potential shifts in climate across important **tipping points** or thresholds, such as those associated with changing ocean circulation or ice sheet collapse (Lenton *et al.*, 2008). Distinguishing between these types of changes is important because different adaptation responses may be appropriate across different timescales. For example, in the Mekong Delta, responding to flood risks from increasingly intense rainfall in the short term might involve riverbank flood-protection measures, but a more appropriate adaptation measure over the longer term might be to change the land use or depopulate the area.

A key challenge has been to link climate change with the aspects of development most pertinent to policy making and the lives of poor people in developing countries (Challinor, 2008; Wilby *et al.*, 2009). For example, in Caribbean islands looking to improve cyclone preparedness, standard rainfall or temperature outputs from climate models may not be of much use. For farmers in semi-arid lands of East Africa, changes in the timing of rainy seasons or increases in rainfall intensity may be more important than changes in the average amount of rain during the year.

In reality, adaptation is usually a response to a wider set of changes that may include for example changes in prices, conflict, security or markets and globalization (Leichenko and O'Brien, 2008). Responses will usually be geared towards realizing immediate benefits rather than only future benefits. As a result: 'Adaptation measures are likely to be implemented only if they are consistent with or integrated with decisions or programs that address non-climatic stresses' (Smit *et al.*, 2001: p. 879).

1.2 Who or what adapts?

Clarity is also needed in defining what is being subjected to impacts and adaptation responses. This may be a natural system (e.g. a coastal ecosystem), or a human system (e.g. a household, community, region, city or particular system of infrastructure), or inter-related human and natural systems (e.g. a mangrove that provides a buffer against sea storms, or a watershed that provides water purification and supply services for residents of a city). Systems of different sizes may require very different adaptation responses:

> ... adaptation at the level of a farmer's field might involve planting a new hybrid; at the farm level it might involve diversification or taking out insurance; at the regional or national scales adaptation may relate to changes in the number of farmers or modifications to a compensation programme; and at a global level, it may involve a shift in patterns of international food trade (Smit *et al.*, 2000: pp. 235–6).

As well as size, understanding the characteristics of the system in question is crucial to determining impacts and adaptation measures. Climate change assessments tend to describe these characteristics in terms of vulnerability, sensitivity, exposure, adaptive capacity and resilience (see Table 5.1).

Table 5.1 *Commonly used terms to describe system characteristics with regards to climate stimuli*

Vulnerability	The propensity or predisposition to be adversely affected.
Sensitivity	The degree to which a system is affected, either adversely or beneficially, by climate-related *stimuli*.
Exposure	The presence of people; livelihoods; environmental services and resources; infrastructure; or economic, social or cultural assets in places that could be adversely affected.
Adaptive capacity	The combination of the strengths, attributes, and resources available to an individual, community, society, or organization that can be used to prepare for and undertake actions to reduce adverse impacts, moderate harm, or exploit beneficial opportunities.
Resilience	The ability of a system and its component parts to anticipate, absorb, accommodate, or recover from the effects of a hazardous event in a timely and efficient manner, including through ensuring the preservation, restoration, or improvement of its essential basic structures and functions.

Source: IPCC, 2012.

1.3 How does adaptation occur?

Adaptation may be a 'reactive' response to an actual climate stress or shock, or a 'proactive' response in anticipation of future climate change. Adaptation also varies according to who is taking the action, commonly divided into responses that are 'planned', usually by governments, and those that occur largely through 'autonomous', private or self-directed efforts (see Table 5.2 for examples).

The vast majority of adaptive actions will likely be autonomous (CCCD, 2010). Nevertheless, governments are responsible for adaptation of public assets (such as forests or hospital buildings) and services (such as provision of clean water or health care) that are sensitive to climate change impacts. Governments also influence the autonomous adaptation responses of others through policies, regulations and **public goods**, for example through incentives for efficient water use, building regulations, weather forecasts, early warning systems, or agricultural extension services (OCED, 2009; CCCD, 2010).

Asking how adaptation occurs also highlights the distinction between coping strategies that usually serve to reduce impacts and maintain the current status, and adaptation, which has a more forward-looking, long-term dimension (see Box 5.1).

Table 5.2 *Dimensions of adaptation*

		Anticipatory/proactive	Reactive
Natural systems		Not applicable	• Changes in length of growing season • Changes in ecosystem composition • Wetland migration
Human systems	*Private/autonomous*	• Purchase of insurance • Construction of house on stilts • Redesign of private infrastructure	• Changes in farm practices • Changes in insurance premiums • Purchase of air-conditioning
	Planned/Public	• Early warning systems • New building codes, design standards, land-use regulations • Incentives for relocation	• Compensatory payments, subsidies • Enforcement of building codes

Source: IPCC, 2001, adapted from Klein, 1998.

Box 5.1

How is adaptation different from coping?

The terms adaptation and coping are sometimes used interchangeably, leading to confusion about the similarities and differences between these two important concepts. The following lists of characteristics are a compilation of brainstorming sessions by groups of development practitioners in Ghana, Niger and Nepal.

Coping

- Short-term and immediate
- Oriented towards survival
- Not continuous
- Motivated by crisis, reactive
- Often degrades resource base
- Prompted by a lack of alternatives

Adaptation

- Oriented towards longer term livelihoods security
- A continuous process
- Results are sustained
- Uses resources efficiently and sustainably
- Involves planning
- Combines old and new strategies and knowledge
- Focused on finding alternatives

Source: Dazé *et al.*, 2009.

1.4 How good is the adaptation?

Finally, an evaluation of adaptation efforts must be based on a set of criteria against which success can be judged. Defining a single set of criteria is complicated due to the multifaceted nature of adaptation and the wide range of adaptation options and contexts. As a result, a number of experts have attempted to highlight broader normative principles on which to evaluate successful adaptation (see for example

Hedger *et al.*, 2008; Wilby and Dessai, 2010; Brooks *et al.*, 2011).
These include:

- *Effectiveness* in achieving stated objectives.
- *Robustness* such that the response is appropriate for a range of possible future conditions.
- *Flexibility* to respond to changing conditions.
- *'Low-regrets'* i.e. leading to outcomes that provide benefits irrespective of climate impacts.
- *Cost-effectiveness* of the response compared to other options.
- *Equity* such that actions reduce or at least do not reinforce existing inequalities between communities, sectors or regions.
- *Legitimacy* such that decisions are accepted by those affected.

Monitoring and evaluation of adaptation is challenging, not least because it is often closely entangled with development interventions (see Box 5.2). There is growing acknowledgement that such evaluation needs to focus on the process of adaptation rather than adaptation as a set of defined end-points. This highlights the need for a learning approach (Tschakert and Dietrich, 2010; Tanner *et al.*, 2012), learning from unsuccessful as well as successful adaptation, especially as different definitions of success are likely to exist among different groups of people. Unsuccessful adaptation may mean that an adaptive action did not deliver the expected benefits, but it may also result in what is known as **maladaptation**.

Maladaptation can be defined as 'action taken ostensibly to avoid or reduce vulnerability to climate change that impacts adversely on, or increases the vulnerability of other systems, sectors or social groups' (Barnett and O'Neill, 2010: 211). Using an example of water management in Australia, Barnett and O'Neill highlight five pathways through which maladaptation arises, characterized by actions that, relative to alternatives:

1   Increase emissions of greenhouse gases (e.g. through pipeline construction and large-scale desalination plants).
2   Disproportionately burden the most vulnerable (e.g. through higher water prices as poorer people spend a higher share of their incomes on water).

3   Have high opportunity costs (e.g. high social and environment costs of pipelines or desalination plants).

4   Reduce incentives to adapt (e.g. creating a change in social norm away from responsible water conservation back to increased consumption).

5   Set paths that limit future choices (e.g. committing capital and institutions to trajectories that are difficult to change in the future).

Box 5.2

The challenges of evaluating successful adaptation

Adaptation provides a number of challenges for conventional approaches to monitoring and evaluation:

- Success criteria: There is no single definition of success, so evaluation is judged against criteria defined by each intervention.

- Metrics: Unlike climate change mitigation (which can ultimately be measured in terms of greenhouse gas emissions), there is no single metric to measure adaptation because goals will depend on different contexts.

- Timescales: Different adaptation options will be appropriate for different timescales, and successful adaptation will be judged against the climate changes that actually occur rather than those projected when plans are formulated.

- Spatial scales: Both climate impacts and adaptation measures cut across spatial scales, which may not be captured by focusing only on narrow project goals.

- Vulnerability indicators: Again, these vary widely according to context. They can represent an intermediate step between the evaluation of institutional changes (e.g. processes associated with capacity building), and the evaluation of improved development outcomes (e.g. increased productivity, reduced disaster losses).

- Shifting baselines and attribution: Most adaptation to climate change will by definition take place against a shifting climatic and environmental baseline. This has the potential to act as a confounding factor in the assessment of development and adaptation interventions.

- Avoiding 'maladaptation': Adaptation efforts may inadvertently increase vulnerability to climate change, but indicators for such failure are poorly developed.

Sources: Adger et al., 2004; Brooks et al., 2011; and Silva Villanueva, 2011.

2. Adaptation in developing countries

2.1 The importance of adaptation for developing countries

After initially being overshadowed by mitigation concerns, adaptation has risen up the international climate agenda and is now a core pillar of the international climate regime. This reflects developing countries' concerns and the widespread acknowledgement that adaptation is a necessity regardless of the success of mitigation efforts because of the lag between emissions and impacts (see Chapters 1 and 2).

As examined in Chapter 2, adaptation is particularly important for poorer countries and poor people within those countries as they are generally the most vulnerable to effects of climate change (see Chapter 2, section 3.2). The conditions and circumstances of those living in poverty in developing countries not only tend to exacerbate their vulnerability to climate change impacts, but also constrain their ability to cope with and adapt to those impacts. This makes adaptation vital in order to limit the negative impacts of climate change on development progress and poverty reduction, including achievement of the Millennium Development Goals (MDGs).

Nevertheless, although poverty is linked to vulnerability, distinguishing between the two is important. Not every adaptation intervention will necessarily reduce poverty. Similarly, the process of poverty reduction may actually increase vulnerability, for example in the case where human settlement or agricultural investments occur in areas that may become unsuitable in future through sea level rise or higher exposure to infectious diseases or heat stress (OECD, 2009). In addition, although poor people's vulnerability is often used to justify adaptation, many measures still do not explicitly address impacts on poor people or target the links between poverty and vulnerability (Eriksen et al., 2011).

2.2 Adaptive capacity and developing countries

The potential, capability, or ability of a system to adapt to the effects of climate change is known as adaptive capacity (IPCC, 2001a) and is intricately tied up with other aspects of development. The level of adaptive capacity tends to be lower for poorer people and societies (see Chapter 2, section 3.3). Some elements of adaptive capacity may be generic, such as education, income and health, while others are

Plate 5.1 *Floating gardens in the Wetland Research and Training Centre (WRTC) run by the Bangladesh Center for Advanced Studies (BCAS).*

Credit: WRI /Aarjan Dixit.

Note: In Bangladesh, 'baira' or floating beds of water hyacinth can be used for growing vegetables. These are being adopted as an adaptation strategy in areas prone to flooding and waterlogging.

specific to different climate impact or development contexts, such as drought or floods, or types of technology (Adger *et al.*, 2007). The determinants of adaptive capacity include (IPCC, 2001a):

- Availability and access to **Economic Resources**.
- Availability and access to **Technology** (e.g. warning systems, protective structures, crop breeding and irrigation, settlement and relocation or redesign, flood control measures).
- **Equity** in the allocation of power and access to resources.
- The structure and organization of critical **institutions**, and access to and participation in decision-making processes.
- **Human capital, skills** and the ability to access and manage **information** for decision-making, including on climate variability and change.
- **Social cohesion** (at societal level) or **social capital/ networks** (at individual level).

For many developing countries, geographical exposure to climate **hazards** and the sensitivity of livelihoods systems are compounded by deficiencies in adaptive capacity. There is often limited information available for adaptation decision-making (including poor climate information from meteorological records, limited dissemination, and low scientific understanding of climate change and its impacts on development), weak institutions and governance (including unclear mandates, weak planning systems, un-inclusive service delivery, and a lack of accountability or participation in decision-making), and lower resources with which to respond (financial resources, skills and experience).

Income levels and human capital are critical generic determinants of adaptive capacity. Better-off households can resort to a range of damage minimization, coping and recovery strategies in the face of external shocks and stresses, including by drawing down savings, claiming insurance for damages incurred, borrowing money, or trading non-essential assets. By contrast, poor people have a more limited scope to access coping strategies or reduce consumption without damaging their health, security, well-being and future development prospects. Indeed, coping with climate shocks and stresses may involve a move to lower-risk and lower-return activities, such as planting subsistence cassava instead of a cash crop, which may perversely reinforce conditions of poverty (UNDP, 2007; see Box 2.5 in Chapter 2).

At a broader level, national and regional economies help to shape both vulnerability and adaptive capacities. Strong economies support the livelihood opportunities poor people need to lift themselves out of poverty and provide the means to finance important public services such as health and education, which are key to **human development**. At the same time, 'some growth strategies can also add to vulnerability such as when they are heavily dependent on one activity, such as tourism, or entail the large-scale depletion and degradation of the natural resource base on which the poor depend disproportionately for their livelihoods' (World Bank, 2010a).

Social cohesion and levels of inequality are also important determinants of vulnerability and adaptive capacity. Social cohesion is linked with the social institutions such as kinship systems, civic organizations, religious groups and other strategies that help mediate shocks (Jenson, 2010). Countries with higher levels of income inequality have been shown to experience the effects of climate

Box 5.3

Vulnerability, adaptation and gender

Vulnerability and adaptive capacity are socially differentiated along lines of age, ethnicity, class, religion and gender. There are structural differences between men and women through, for example, gender-specific roles in society, work and domestic life. These differences affect the vulnerability and capacity of women and men to adapt to climate change.

In the developing world in particular, women are disproportionately involved in natural resource-dependent activities, such as agriculture, compared to salaried occupations. As these activities are directly dependent on climatic conditions, changes in climate variability projected for future climates are likely to affect women through a variety of mechanisms: directly through the availability of water, vegetation and fuel wood, and through health issues relating to vulnerable populations (especially dependent children and the elderly). Water scarcity and declining rainfall levels may force women to walk longer distances to have access to water and fuel wood, and thus reduce the amount of time they can spend cultivating their fields (resulting in lower yields) and/or earning money through a variety of income-generating activities.

Most fundamentally, the vulnerability of women in agricultural economies is affected by their relative insecurity of access to rights over information, knowledge, resources and sources of wealth such as agricultural land. For example, women have often been neglected by agricultural extension officers, therefore limiting their access to new crop varieties and agricultural technologies. In addition, it is well established that women, in many countries, are disadvantaged in terms of property rights and security of tenure, though the mechanisms and exact form of the insecurity are contested. This insecurity can have implications both for their vulnerability in a changing climate, and for their capacity to adapt productive livelihoods to a changing climate.

(Source: OECD, 2009, adapted from Adger et al., 2007).

disasters more profoundly than more equal societies (Roberts and Parks, 2007). These inequalities in vulnerability and adaptive capacity also reflect a range of social factors, including ethnicity, age and gender (see Box 5.3).

2.3 Closing the adaptation deficit and the 'low-regrets' approach

Deficiencies in adaptive capacity are reflected in many places by the limited ability to manage and respond to existing variability of the climate, even before additional climate change is taken into account. In such cases, improving the capacity of communities, companies and governments to deal with climate variability can also build the capacity to tackle future changes in the climate. For example, managing water resources more efficiently and equitably is crucial because of the current poor management and the increasing demands for water in many regions, regardless of the impacts of climate change on water availability. By tackling these existing challenges, water managers will be in a much better position to plan for a changing climate (Ludwig *et al.*, 2009; Agnew and Woodhouse, 2010). Such options, which deliver development benefits at the same time as enhancing adaptive capacity, are also referred to as win-win options, with those that also contribute to mitigation known as win-win-win.

Closing this pre-existing **adaptation deficit** has therefore been proposed as a baseline adaptation strategy for developing countries, as they respond to existing development priorities (Burton and May, 2004; Kok *et al.*, 2008; Heltberg *et al.*, 2010; Wilby and Desai, 2010). Measures to deal with the adaptation deficit are referred to as **no-regret** options because they are not affected by uncertainties related to future climate change. For example, improved provision and dissemination of early warning systems will be beneficial irrespective of how the climate changes. Some related adaptation measures are referred to as **low-regret** options as they require some investment directed at tackling future climate change, but should be likely to deliver large benefits under relatively low risks. For example, wider drains may be required in order to deal with increases in rainfall intensity. Installing these at the time of planned investment in new drain construction or improvement is likely to be cheaper than having to retro-fit at a later date.

2.4 Adaptation and Disaster Risk Reduction

Many of the options for closing the adaptation deficit by building resilience to existing climate variability draw from the policies and practices of **disaster risk reduction (DRR)**. DRR is defined as: Action taken to reduce the risk of disasters and the adverse impacts of natural hazards (UNISDR, 2008). It denotes both a policy goal or objective, and the strategic and instrumental measures employed for anticipating future disaster risk; reducing existing exposure, hazard, or vulnerability; and improving resilience (IPCC, 2012).

DRR represents a change in emphasis, away from humanitarian response and rehabilitation following disaster events, and towards reduction of potential risks and prevention of disasters themselves.

This shift has focused greater attention on the social, economic and political factors that influence social vulnerability, rather than just the nature of the hazard and local environmental conditions (Schipper and Pelling, 2006; Wisner *et al.*, 2004). DRR is therefore seen as situated within broader development processes and disasters can be regarded as resulting from development failure, 'where the root causes of vulnerability merge with the origins of other development-related crises' (Gaillard, 2010: p. 222).

DRR can be seen as the first line of defense against climate change impacts, such as increased flooding or regular droughts, and has become part of the standard tool kit for climate change adaptation programmes (Mitchell and van Aalst, 2008). At the same time, the DRR community is increasingly concerned with the influence of climate in changing the burden of hazards such as rainfall, temperature or cyclones (see Chapter 1, section 5 on the IPCC SREX report).

The Hyogo Framework for Action, agreed in 2005 with the support of 168 governments, provides the foundation for the international implementation of disaster risk reduction. It sets out five priorities for action, each elaborated into a number of specific areas of attention. These offer a strong basis for developing risk-reducing adaptation measures:

1 Ensure that disaster risk reduction is a national and local priority with a strong institutional basis for implementation.
2 Identify, assess and monitor disaster risks and enhance early warning.
3 Use knowledge, innovation and education to build a culture of safety and resilience at all levels.
4 Reduce the underlying risk factors.
5 Strengthen disaster preparedness for effective response at all levels.

Activities supported by Oxfam in the Indian state of Andhra Pradesh, whose coastline is vulnerable to cyclones, provide an example of the wide range of DRR responses (Oxfam GB, India, 2004). Measures include development of village contingency plans, food security schemes in which a handful of rice at every meal is to be stored in a safe place and used during the cyclone season, and training multipurpose health workers to carry out regular health checks and deal with emergency problems such as snakebites and waterborne diseases in the wake of a cyclone. Physical measures have included retro-fitting the houses of those most vulnerable to cyclone damage and planting

Casuarina trees, which 'bend and swing, but they will not break'. Income-generation schemes, often focused on households headed by women, have included a fish-drying project, support to buy and run cargo boats (which are also used for emergency relief), and businesses producing low-cost, cyclone-resistant bricks.

3. Analytical approaches to climate change adaptation

Extending other analyses, we introduce four different analytical approaches to adaptation, with their characteristics summarized in Table 5.3 (Burton *et al.*, 2002; Smit and Wandel, 2006; Eakin and Patt, 2011). This typology of approaches helps to illustrate and emphasize differences in approach, tools and methods. One commonality is the logical sequence that underpins most approaches to adaptation that includes:

- Defining the problem and selecting the methodology.
- Assessing the impacts against existing adaptation and development processes.
- Selecting and implementing additional adaptation strategies.
- Evaluating the performance of strategies.

Table 5.3 *Typology of different approaches to adaptation*

Approach	Impacts-led	Vulnerability-led	Resilience	Policy and decision-making
Dominant methods used?	Climate science and impact models	Vulnerability and capacity assessments	Systems-based assessments	Policy and decision-making analysis
Adaptation to what?	Future, gradual climate change	Existing climate variability, extremes	Changing climate variability, uncertainty	Climate variability and change
Main focus of who or what adapts?	Ecosystems, infrastructure, technology	Livelihoods, social organisation	Institutions, socio-ecological systems	Policy and decision-making systems
How does adaptation occur?	Discrete adaptation to climate change	Reducing vulnerability and enhancing adaptive capacity	Ability to bounce back from shocks and stresses, learning	Improved decision-making and learning processes
How good is the adaptation?	Adaptation evaluated against no adaptation scenario	Adaptation evaluated against elements of adaptive capacity	Performance of different parts of the socio-ecological system	Performance against adaptive decision-making criteria

3.1 Impacts-led approaches

Early adaptation discourses were framed by global scientific assessments and marked by an orientation towards engineering solutions. Early research and analysis tended to emphasize the impacts of climate change and give primacy to the role of climate science and modelling in understanding impacts and adaptation. They were underpinned by standardized methodologies developed under the IPCC in the early 1990s (Carter *et al.*, 1994). In applying climate change scenarios, they emphasize the future conditions rather than current impacts or vulnerability (Burton *et al.*, 2002). Such impact-based assessments are particularly important in highlighting regional and global implications of climate change, such as those around water and food security (see Box 5.4). These provide compelling evidence to stimulate both mitigation and adaptation action.

Box 5.4

Global food production and climate change

Parry *et al.* (2004) analyze the global consequences to crop yields, production, and risk of hunger of linked socio-economic and climate change scenarios. Generally, the SRES greenhouse gas emissions scenarios (see Box 1.4 in Chapter 1) result in crop yield decreases in developing countries and yield increases in developed countries.

The scenario with the largest greenhouse gas increases (A1FI), as expected with its large increase in global temperatures, exhibits the greatest decreases both regionally and globally in yields, especially by the 2080s. Decreases are especially significant in Africa and parts of Asia with expected losses up to 30 per cent. In these locations, effects of temperature and precipitation changes on crop yields outweigh the beneficial direct fertilisation effects of greater CO_2 concentrations in the atmosphere.

When crop yield results are combined with a world food trade system model, the combined model and scenario experiments demonstrate that the world, for the most part, appears to be able to continue to feed itself under all the SRES scenarios during the rest of this century. However, this outcome can only be achieved through production in the developed countries (which mostly benefit from climate change) compensating for declines projected, for the most part, for developing nations. While global production appears stable, regional differences in crop production are likely to grow stronger through time, leading to increasing regional differences in impacts, with substantial increases in prices and risk of hunger amongst the poorer nations, especially under scenarios of greater inequality (A1FI and A2).

Source: Parry *et al.*, 2004.

Box 5.5

Climate proofing infrastructure in the Pacific Islands

Climate proofing infrastructure at the design stage is seen as contributing to the sustainability of development investment, as it extends its life in the face of future climate risk. In Rarotonga, the Cook Islands, the design of the breakwater for the western basin of the new Avatiu Harbour was climate-proofed. The breakwater was under redesign to expand the harbour's vessel capacity, required due to tuna fishing activity growth.

To climate proof this investment, model data was used to take into account climate change scenarios for cyclone intensity (translated into changes in wave heights affected by wind speed) and sea level rise (as a component of the change in water elevation during cyclones). Changes in cyclone activity and projected sea level changes suggested a reduction by half of the return periods (i.e. an increase in the occurrence) of extreme wind speeds by 2050 and an increase in the once-every-50-years extreme wave height from 10.8 m with current climate to 12m with climate projections. Given these results, the design of the harbour breakwater protection was increased to cope with wave heights of 12 m and at least a 50 cm sea level rise.

Source: ADB, 2005.

The use of impact models to frame adaptation responses continues to drive engineering approaches to adaptation associated with the concept of 'climate proofing' infrastructure (see Box 5.5). It also results in a bias towards technology-based adaptation, which lends itself better to quantification and integration into impact modelling. Whilst impact models can help provide general assessments of impacts and cross-comparability across regions, the guidance they provide is of limited use for adaptation at local scales due to high levels of uncertainty. These approaches also tend to underplay the social dimensions of vulnerability and adaptation, thereby narrowing the range of adaptation options considered. For example, an assessment of agriculture based on a crop yield model might consider new irrigation systems or introduction of pesticides or crop hybrids, rather than 'softer' management-based options, land tenure or off-farm market conditions (Burton *et al.* 2002).

3.2 Vulnerability-led approaches

In contrast to the impacts-led assessments, vulnerability-led assessments place more emphasis on factors internal to the affected system and how they form the 'starting-point' conditions for dealing with shocks and stresses (O'Brien *et al.*, 2007). They place greater emphasis on social aspects such as livelihoods, institutions and information, and rather than focusing on future conditions, they stress current vulnerability conditions and the management of existing climate variability.

Vulnerability-based approaches were developed partly as a result of work on adaptation in contexts where planning horizons were short and decision makers faced significant uncertainty in model projections. The starting points for these approaches are historic and current vulnerability and coping strategies, which are used as the baseline onto which future climate change impacts and adaptation activities are overlaid. This stresses the inseparability of the climate context from the changing development context that affects vulnerability and adaptive capacity, leading some commentators to note that much adaptation to climate change is simply an extension of good development practice (McGray *et al.*, 2007; Stern, 2007).

Adaptation actions in these approaches do not start from scratch but are based on existing institutions, policies and practices and are part of a response to multiple shocks and stresses, rather than only to climate stressors (Burton *et al.*, 2002). For example, improving adaptive capacity through peace-building activities may be the most pressing concern in conflict-affected areas. Such actions may not even address climate risks directly, but rather be directed at improving people's ability to cope with wider adversity and change, such as through social protection (see Table 5.4), assisted migration (see Box 5.6), or access to microfinance (Hammill *et al.*, 2008; Eakin and Patt, 2011).

By shifting attention away from impacts, these approaches emphasize the differences in impacts, vulnerability and adaptive capacity for groups of people and contexts, and highlight the links between poverty and vulnerability (Eriksen and O'Brien, 2007). This focuses analysis on the diversity of adaptation options required for different contexts for different people, places and timescales (Tanner and Mitchell, 2008). Placing social aspects of vulnerability centre stage has also highlighted the need to target and actively involve vulnerable and marginalized groups such as women and children to ensure their needs are understood and met (Demetriades and Esplen, 2008; Tanner *et al.*, 2009; see Box 5.3 earlier in the chapter).

The vulnerability approach has led to integration of climate change with other approaches designed to tackle livelihood risks, such as

Table 5.4 *Adaptive Social Protection measures*

Social protection measure	Benefits for adaptation and DRR	Challenges
Weather-based crop insurance schemes	– Rapid payouts possible – Guards against adverse selection and moral hazard – Frees up assets for investment in adaptive capacity – Easily linked to trends and projections for climate change – Supports adaptive flexibility and risk taking	– Targeting marginal farmers – Tackling differentiated gender impacts – Affordable premiums for poor – Subsidizing capital costs – Integrating climate change projections into financial risk assessment – Guarantee mechanisms for reinsurance
Seed transfer schemes	– Boost agricultural production and household food security – Post-disaster response tool – Seed varieties can be tailored to changing local environmental conditions – Cost effectiveness of seed voucher and fair projects – Fairs promote crop diversity and information sharing	– Ensuring locally appropriate seed and fertilizer varieties – Protection of crop diversity – Reduce distortion of local markets – Focus on access rather than only availability – Inclusive approach that draws in marginal farmers
Asset transfer schemes	– Ability to target most vulnerable people – Easily integrated in livelihoods programmes	– Ensuring local appropriateness of assets – Integrating changing nature of environmental stresses in asset selection
Cash transfer schemes	– Targeting those most vulnerable to climate shocks – Smoothing consumption allowing adaptive risk taking and investment – Flexibility enhanced to cope with climate shocks	– Ensuring adequate size and predictability of transfers – Long term focus to reduce risk over extended timeframes – Demonstrating economic case for cash transfers related to climate shocks – Use of socio-ecological vulnerability indices for targeting

Source: Davies *et al.*, 2009.

social protection. Social protection is a term applied to initiatives that transfer income or assets to the poor, protect the vulnerable against livelihood risks, and enhance the social status and rights of the marginalized (Devereux and Sabates-Wheeler, 2004; Davies *et al.*, 2009; see Table 5.4).

Vulnerability-led approaches have heavily influenced development planning and programming, such as through the UNDP-GEF's Adaptation Policy Framework (UNDP, 2005) and the National Adaptation Programmes of Action (NAPA) approach under the UNFCCC (see Chapter 3). They have been particularly popular with NGOs working at community level, where the practice of

Box 5.6

Climate change and migration

Responding to growing reports citing projected numbers of people displaced by climate change in the future, a UK government Foresight report in 2011 sought to investigate the evidence underpinning linkages between migration and global environment change.

It concluded that the impact of environmental change on migration will increase in the future. In particular, environmental change may threaten people's livelihoods, and a traditional response is to migrate. Environmental change will also alter populations' exposure to natural hazards, and migration is, in many cases, the *only* response to this. For example, 17 million people were displaced by natural hazards in 2009 and 42 million in 2010 (note that this number also includes those displaced by geophysical events such as earthquakes).

However, the relationship between climate change and migration is not straightforward:

● Powerful economic, political and social drivers mean that migration is likely to continue regardless of environmental change.

● Environmental change is equally likely to make migration less possible as more probable because migration is expensive and requires forms of capital, yet populations who experience the impacts of environmental change may see a reduction in the very capital required to enable a move.

● Consequently, in the decades ahead, millions of people will be *unable* to move away from locations in which they are extremely vulnerable to environmental change. To the international community, this 'trapped' population is likely to represent just as important a policy concern as those who do migrate.

The report concludes that migration can also be part of the solution in the face of global environmental change. Planned and facilitated approaches to human migration can ease people out of situations of vulnerability. Doing so requires linkages with emerging financing mechanisms as well as building sustainable, flexible and inclusive infrastructure and service delivery in cities, which face the twin challenges of population growth and environmental change.

Source: The Government Office for Science, 2011.

Community-Based Adaptation (CBA) has emerged (Ayers *et al.*, 2013). CBA is a bottom-up approach where community members identify their concerns regarding adaptation, and people's livelihoods and vulnerability are at the centre of the analysis (Dazé *et al.*, 2009). International NGO CARE has developed four inter-related strategies for its CBA work that include:

1 Promotion of climate resilient livelihood strategies in combination with income diversification and capacity building for planning and improved risk management.
2 Disaster risk reduction strategies to reduce the impact of hazards, particularly on vulnerable households and individuals.
3 Capacity development for local civil society and governmental institutions so that they can provide better support to communities, households and individuals in their adaptation efforts.
4 Advocacy and social mobilization to address the underlying causes of vulnerability, such as poor governance, lack of control over resources, or limited access to basic services.

CBA toolkits such as CARE International's Community Vulnerability and Capacity Assessment (CVCA) use participatory methods at the community level to assess current hazards, coping strategies and adaptive capacity, combining these with an analysis of local to national institutional and policy contexts (Dazé et al., 2009). Such assessments have drawn heavily on the participatory tools and methods associated with community development, such as seasonal calendars, hazard mapping, historical timelines and hazard impact matrices (van Aalst et al., 2008; Magee, 2012). Such an approach can bring together scientific information with local and indigenous knowledge to provide a better understanding of the basis from which to shift from existing coping measures to adaptation to future climate change.

One example of CBA uses asset restocking to support adaptation (Davies et al., 2009). Poor people often rely on the sale of assets as a coping strategy for a range of risks including those induced by climate change. The Reducing Vulnerability to Climate Change (RVCC) programme attempted to build the assets of vulnerable communities in south-western Bangladesh with a view to making them better adapted to a changing climate. This programme entailed a number of components such as encouraging communities to move away from livelihoods that could be adversely affected by changing climate impacts.

The project also spread awareness on the construction and use of 'Baira' or floating gardens (see Plate 5.1). These are floating beds of water hyacinth that are prepared on water bodies and can be used for growing vegetables. As opposed to traditional sites of agricultural production in Bangladesh that lie along rivers, these floating gardens are less vulnerable to flooding and provide a source of livelihoods and asset accumulation that is better adapted to climate change. Similarly, the programme sought to make more productive use of waterlogged land by encouraging duck farming. The birds could be sold in times of financial stress or consumed as supplementary nutrition and could be easily reared in areas that had otherwise become useless to more traditional forms of agriculture. Moreover, the ducks, unlike other forms of poultry, were far more resilient to different intensities of flooding – an important impact of climate change in the region.

3.3 Resilience approaches

The concept of resilience has been increasingly explored in the context of climate change. It has become particularly popular in bringing together a range of overlapping issue areas that include adaptation, disaster risk reduction, poverty reduction, food security, nutrition and conflict. Resilience thinking covers a wide range of characteristics (see Box 5.7) and the term 'climate resilient development' is rapidly becoming a catch-all for tackling climate change impacts in a

---

Box 5.7

Characteristics of resilience thinking

Reviewing resilience and adaptation research, Bahadur and colleagues (2013) distil key characteristics of resilience of particular relevance for adaptation that include:

- **A high level of diversity** in terms of adaptation options, livelihood strategies and opportunities, access to assets, engagement of diverse members of the community, and drawing on diverse sources of knowledge in making decisions.

- **Effective institutions that are connected across scales**, able to facilitate learning processes and perform specialised functions such as translating scientific climate data for policy making.

- **Embracing uncertainty and change** rather than resisting it, including by building in redundancy within systems such that partial failure does not lead to the whole system collapsing, and rejecting the idea of restoring systems to a prior stable equilibrium after a disturbance.

- **A high degree of equity**, both social and economic, enabling resilient systems to distribute risks fairly across different parts of the system or community.

Source: Bahadur et al., 2013.

---

development context. While it is rooted in a wide range of disciplines (Bahadur *et al.*, 2013), resilience theories are most closely associated with work on socio-ecological systems, wherein a resilient system is characterized as one that is able to retain core structures and functions in the face of significant disturbance, while still retaining options to develop (Nelson *et al.*, 2007).

The resilience concept has proved popular as a way of communicating the ideas of coping strategies (bouncing back) and adaptation strategies (bouncing back better). This popularity stems in part from resilience as a concept that can draw together actions to respond to a wider range of shocks and stresses, including food or oil price shocks, financial or political crises, or conflicts (see Box 5.8). 'The aim of resilience programming is, therefore, to ensure that shocks and stresses, whether individually or in combination, do not lead to a long-term downturn in development progress' (Mitchell and Harris, 2012: p. 1).

Janssen and Ostrom (2006: p. 238) argue that the resilience approach, '. . . emphasises non-linear dynamics, thresholds, uncertainty, and surprise,' while adaptation focuses either on reducing expected damage from climate scenarios or on '. . . risks that are already problematic'. Resilience therefore focuses on readiness for surprise and brings an understanding of how a system can be organized around multiple stable states as opposed to one equilibrium point. This greatly expands the array of adaptation actions that may be undertaken (Nelson *et al.*, 2007).

Resilience thinking tackles adaptation at the intersection of multiple and overlapping systems. For example, efforts to enhance resilience may require understanding how roads or electricity infrastructure systems relate to water management systems and how that in turn affects small-scale farmers' ability to adapt to a changing climate (ISET, 2008).

Resilience thinking also emphasizes processes of social and organizational learning (Collins and Ison, 2009; Nilsson and Swartling, 2009; Tanner *et al.*, 2012). Learning is central to the notion of adaptive management, which entails considering a range of plausible future changes in the system, weighing up a range of possible strategies against these potential futures, and then favoring actions that are robust in the face of future uncertainties (Gunderson and Holling, 2001; Wilby and Dessai, 2010).

Box 5.8

Climate change and conflict

Does climate change increase the risk of violent conflict? Climate change could reduce access to natural resources and undermine state capacity to help people sustain livelihoods. These impacts may in certain circumstances increase the risk of violent conflict, but further investigation is needed.

Climate change poses risks to human security both through changes in mean conditions and the severity and frequency of natural disasters. Impacts on human systems are worrying given the rate of change and the fact that climatic variations have triggered large-scale social disruptions in the past. Yet environmental change does not undermine human security in isolation from social factors such as poverty, state support, access to economic opportunity and social cohesion. These factors determine people's capacity to adapt to climate change. The way climate change undermines human security varies across the world.

There are four key factors affecting processes by which climate change could exacerbate violent conflict. These are:

- **Vulnerable livelihoods**: Impacts on livelihoods will be more significant for those with high resource dependency and in socially and environmentally marginalized areas.

- **Poverty**: Climate change may directly increase poverty by undermining access to natural resources. It may indirectly increase poverty through its effect on resource sectors and the ability of governments to provide social safety nets.

- **Weak states**: The impacts of climate change are likely to increase the costs of providing public infrastructure and services and may decrease government revenue. Climate change may decrease the ability of states to create opportunities and provide freedoms for citizens.

- **Migration**: People whose livelihoods are undermined by climate change may migrate, though climate change is unlikely to be the only 'push' factor. Large-scale migrations may increase the risk of conflict in host communities.

While climate change poses clear threats, more research is needed on how it may undermine human security. Current understanding is not sufficient for designing adaptation strategies. There is a need for systematic, comparative and cross-scale research on the connections between climate change, human security and violence.

Source: GSDRC summary of Barnett and Adger, 2007. Available at: http://www.gsdrc.org/go/display&type=Document&id=3409

> . . . resilience means more than just responding to, and bouncing back after, an extreme event. It also involves the capacity to change and adapt to changing environmental conditions, and that, in turn, requires the essential abilities to cooperate, learn, and apply the lessons toward continued resilience under future conditions.
>
> (Moser, 2008: p. 17)

Continuous learning is particularly important in responding to climate change (Tanner *et al.*, 2012: p. 10) because:

- we have limited experience with the required responses (so less certainty of achieving results);
- there are competing visions of the problem and its solution (each telling plausible but conflicting tales of climate change);
- climate change introduces new sources of uncertainty due to future emissions scenarios, climate and impact models and the existing uncertainty of the changing development landscape on which impacts play out;
- the cross-sectoral nature of climate change adds an additional layer of complexity (climate change cuts across sectors and scales).

Use of a resilience approach prompts a number of questions. In focusing on the ability to recover from shocks and stresses, resilience thinking may fail to alter the status quo (Pelling, 2010). This contrasts with the transformational shifts in governance and institutions that may be required when incremental adjustments to the existing system have reached their limit (see Chapter 8). Other important considerations stem from the lack of a normative framework for resilience that could answer fundamental questions such as: Managing resilience of what and for whom?; Is resilience always a desirable goal (e.g. the resilience of fossil-fuel based energy systems)?; How do values, power and politics inform resilience processes?; What are the limits of resilience and when is transformation of the system necessary?; How to manage deliberate transformations?; How to move from learning after major events towards more anticipatory approaches? (Leach, 2008; Béné *et al.*, 2012; O'Brien, 2012)

Plate 5.2 *Interviews on environmental change in coastal Ghana for climate change related community radio broadcasts.*

Credit: Blane Harvey.

Note: Despite the rise of more modern information communication technology, radio still provides an important medium for communicating climate change issues in many remote locations.

3.4 Policy- and decision-making centered approaches

As adaptation concerns have been increasingly addressed in the context of development, so there has been increasing attention on analyzing climate impacts, vulnerability and adaptation options from the vantage point of policy implementation and decision-making. As highlighted by the adaptation–development continuum presented in Chapter 2 (Box 2.4), responses to climate change are often inseparable from responses to a wider range of development pressures such as population growth, migration and urbanization, changing trade patterns and globalization, conflict and political instability, or health challenges such as HIV/AIDS (McGray *et al.*, 2007).

Given that any adaptation actions are unlikely to be a response to climate change alone, this approach focuses on the integration, or **mainstreaming**, of climate change adaptation into other areas of development policy and practice (Huq and Reid, 2005; OECD, 2009). There are now a growing number of examples of such integration across a wide range of sectors and contexts, including urban development, natural resource management, community development and livelihoods, coastal zone management, and disaster risk management (Smit and Wandel, 2006; see case studies in McGray *et al.*, 2007, and in WRI, 2011).

At the broadest level, this integration process, often referred to as 'climate risk management', can be seen in terms of applying a 'climate lens' to development interventions (strategies, policies, plans, programmes or regulations) (OECD, 2009: p. 17) in order to examine:

- the extent to which a measure – be it a strategy, policy, plan or programme – under consideration could be vulnerable to risks arising from climate variability and change;
- the extent to which climate change risks have been taken into consideration in the course of the formulation of this measure;
- the extent to which it could increase vulnerability, leading to maladaptation or, conversely, miss important opportunities arising from climate change; and
- for pre-existing strategies, policies, plans and programmes that are being revised, what amendments might be warranted in order to address climate risks and opportunities?

A range of tools has been developed and employed to facilitate this integration of adaptation and development, which can be divided into three types that focus on: (1) data and information generation; (2) knowledge sharing; and (3) process guidance (Hammill and Tanner, 2011; see Table 5.5).

Focusing on policy implementation and decision-making has also led to a greater focus on the criteria to select adaptation options. Rather than analyzing costs and benefits in terms of performance against a certain future outcome, decision-making should take into account the uncertainty inherent in future climate change and future vulnerability. Figure 5.1 illustrates five criteria for

Table 5.5 *Categories and examples of adaptation tools*

....

Type/characteristics	*Examples from the development community*

....

1. Process guidance tools

Tools that guide users through the identification, gathering, and analysis of relevant data and information to: • Identify climate risks to development activities (often using Type 2 tools) • Assess and analyse climate risk management strategies • Evaluate options to integrate climate risk management into development activities	• Adapting to Coastal Climate Change: A Guidebook for Development Planners www.crc.uri.edu/index.php?actid=366 • BMZ Environment and Climate Assessment www.gtz.de/climate-check • CEDRA: http://tilz.tearfund.org/Topics/Environmenal+Sustainability/CEDRA.htm • CRiSTAL: www.cristaltool.org • ORCHID: www.ids.ac.uk/climatechange/orchid • USAID Guidance Manual: http://www.usaid.gov/our_work/environment/climate/policies_prog/adaptation.html

2. Data and information provision tools

Often depending on some computer capacity, these tools generate or present data and information on: • Primary climate variables and projections (e.g. temperature, rainfall trends) • Secondary climate impacts (e.g. flood maps, crop yields) • Vulnerability and response options (e.g. poverty maps, example adaptation options)	• CI-Grasp www.ci-grasp.org • Climate Wizard: www.climatewizard.org • Climate Change Explorer Tool: www.weadapt.org/wiki/The_Climate_Change_Explorer_Tool • PRECIS: www.precis.metoffice.com • SERVIR: www.servir.net • World Bank CC Knowledge Portal: climate, impact and socio-economic data http://sdwebx.worldbank.org/climateportal/

3. Knowledge-sharing tools

Platforms and networks that offer adaptation practitioners a virtual space for information and experiences related to climate risk and adaptation. These spaces allow users to : • House or store information and knowledge • Share it with other interested users • Interact with other users to develop or advance ideas, approaches, tools, monitoring etc.	• Adaptation Learning Mechanism: www.adaptationlearning.net • AfricaAdapt: www.africa-adapt.net • Climate Adaptation Knowledge Exchange: www.cakex.org • ELDIS resource guide on Adaptation: www.eldis.org/go/topics/dossiers/climatechange-adaptation • weADAPT platform: www.weadapt.org • World Bank CC Knowledge Portal: http://sdwebx.worldbank.org/climateportal/

....

Source: Hammill and Tanner, 2011.

effective adaptation decision-making (WRI, 2011) that include:

- A **responsive** decision-making process would advance policies/plans *after* a climate change has occurred, to react quickly to the climate change.
- A **proactive** decision-making process will create policies/plans in advance of a climate change that has yet to occur; the decision-making process prepares for that climate change and its impacts.

- A **flexible** decision-making process adjusts policies/plans based on ongoing climate changes, with each response readjusted due to learning from previous experiences and new conditions on the ground.
- A **durable** decision-making process advances policies/plans that can accommodate the long-term nature of some climate changes.
- A **robust** decision-making process would result in policies/plans that are effective in managing a full range of possible impacts associated with a given climate change; that is necessary due to the uncertainty regarding the timing, scope and scale of some climate changes.

Finally, there is increasing awareness of the need to understand the role of institutions and the political economy of adaptation decision-making in supporting adaptation actions (Tanner and Allouche, 2011). Adaptation responses span the remits of numerous government departments and organizations, posing challenges for institutional coordination, especially in countries where government capacity is already limited. A compartmentalized approach in many countries makes coordination across adaptation, mitigation and development even more challenging (Hagemann *et al.*, 2011). At the same time, the coordination and leadership role for climate change commonly rests with the environment ministry, which is usually poorly funded and lacking in convening power across other government departments. Nevertheless, there are examples of improvements in coordination and

Figure 5.1 *Criteria for effective adaptation decision-making.*

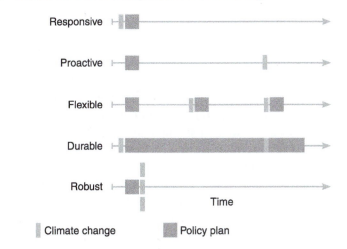

Source: WRI, 2011.

Box 5.9

Policy making and linked institutions: Lessons from flood management in Mozambique

Climate risk management to reduce the severity and impacts of flooding requires coordination among institutions across sectors and scales, from national policies to specific measures at the local level before, during and after the disaster, as well as international information sharing and coordination.

In 2000, Mozambique experienced its most severe flood, a result of a combination of increased rainfall outside and within the country, and cyclone activity. Coordination was led by its flood early warning system, composed and coordinated by the National Disaster Management Institute, the National Directorate of Water, and the National Institute of Meteorology (NIM). Weather forecasts and cyclone monitoring were provided by the NIM. River basin administrations coordinated with district and local government authorities, and with the Red Cross and other NGOs for the flood warnings and evacuations at the local level. While these early warning systems were considered to have worked relatively well, the communication linkages with the general public (such as through media) were weak or non existent. As a result, in learning lessons from the floods, increased efforts for linking the met office with the media were taken, as well as reinforcement of the importance of supranational coordination and climate modelling and monitoring, including a basin-by-basin vulnerability analysis.

Source: Hellmuth *et al.*, 2007.

---

communication regarding climate risk management and adaptation (see Box 5.9).

Conclusion

Cumulative greenhouse gas emissions to date and the ongoing failure to cut emissions at the rate necessary to avoid dangerous levels of climate change mean that adaptation to climate change impacts is an urgent necessity. The need for adaptation processes is particularly acute in poor countries and among poor people, characterized by heightened sensitivity to climate shocks and stresses and weak capacity to respond. The adaptation challenge is often presented as being predominantly about scientific information on climate impacts and access to international finance (see Chapters 1 and 6). This chapter has highlighted the need to consider also the vulnerability and adaptive capacity of different people and systems, how to build longer term resilience, and consideration of decision-making contexts in adapting development processes to climate change. This in turn entails understanding and transforming decision-making behavior and, in doing so, our approach to securing future human welfare in a resource-constrained world (see Chapters 7 and 8).

Summary

- The emergence of adaptation issues in international climate negotiations has been driven largely by developing countries, concerned that their geographical exposure, natural resource-dependent livelihoods, and economic capacity to respond made them particularly vulnerable to climate-related shocks and stresses.

- Adaptation can be unpacked in terms of the change to which development is being adapted, the system that is adapting, how the adaptation happens and evaluating how good the adaptation is.

- Adaptation is heavily interlinked with responses to wider development shocks and stresses, especially in responding to the 'adaptation deficit' through actions that deliver development benefits irrespective of the nature of climate change.

- Approaches to adaptation have evolved, with a range of analytical tools and methods. These can be broken down as a typology of four different approaches to adaptation: Impacts-led, vulnerability-led, resilience, and policy- and decision-making centered.

Discussion questions

- How does adaptation differ from good development practice?

- What are the limitations of impact-focused approaches for adaptation decision-making and how can these limitations be addressed?

- What are the benefits and drawbacks of using the concept of 'resilience' for climate adaptation?

- What are the gender implications of vulnerability and adaptation to climate change and how can these implications be addressed in development initiatives?

- What are the challenges and pitfalls of a tool-based approach to integrating climate change adaptation into development organizations?

Further reading

Ensor, J., Huq, S. and Berger, R. (eds) (2013) *Community-Based Adaptation to Climate Change: Emerging lessons*, Rugby: Practical Action Publishing.

Moser, S.C. and Boykoff, M.T. (eds) (2013) *Successful Adaptation to Climate Change: Linking Science and Practice in a Rapidly Changing World*, Abingdon: Routledge.

Pelling, M. (2010) *Adaptation to Climate Change: From Resilience to Transformation*, Abingdon: Routledge.

Schipper, E.L. and Burton, I. (eds) (2009) *Earthscan Reader on Adaptation to Climate Change*, London: Earthscan.

Tanner, T.M. and Mitchell, T. (eds) (2008) 'Poverty in a Changing Climate', *IDS Bulletin* 38.4.

Tanner, T.M., Lockwood, M. and Seballos, F. (2012) *Learning to Tackle Climate Change*. Brighton: Institute of Development Studies. Available at: www.ids.ac.uk/idspublication/learning-to-tackle-climate-change

World Resources Institute (WRI) in collaboration with United Nations Development Programme, United Nations Environment Programme, and World Bank (2011) *World Resources 2010–2011: Decision Making in a Changing Climate—Adaptation Challenges and Choices*, Washington, DC: WRI.

Websites

www.adaptationlearning.net
The **Adaptation Learning Mechanism** is run by a collective of UN agencies and the World Bank, providing a common platform for sharing and learning on adaptation practices; integrating adaptation into development policy, planning and operations; and capacity building.

www.africa-adapt.net
AfricaAdapt is an independent bilingual network (French/English) focused on Africa. The network's aim is to facilitate the flow of climate change adaptation knowledge for sustainable livelihoods between researchers, policy makers, civil society organizations and communities who are vulnerable to climate variability and change across the continent.

www.eldis.org/go/topics/resource-guides/climate-change/climate-change-adaptation
ELDIS resource guide on climate change adaptation, with key issues guides and
   summaries of key articles updated regularly. Extensive collection on community
   based adaptation.

http://sdwebx.worldbank.org/climateportal/
World Bank Climate Change Knowledge Portal contains environmental, disaster
   risk, and socio-economic datasets, as well as synthesis products, such as the
   Climate Adaptation Country Profiles. The portal also provides intelligent links to
   other resources and tools.

6 Climate finance

The US and Europe alone have spent US$4.1 trillion on rescuing financial firms since the beginning of the global recession in 2008, more than 40 times the money all industrialized countries have pledged to help developing countries tackle climate change by 2020.

      ◉ **Finance for tackling climate change: What, why and how much?**
      ◉ **Understanding the different sources and channels of climate finance**
          ○ **UNFCCC climate finance mechanisms**
          ○ **Bilateral and multi-lateral climate finance**
          ○ **National responses to climate finance**
          ○ **Climate finance and the private sector**
          ○ **Proposals for future climate finance**
      ◉ **A framework for assessing climate finance: Governance, sources of revenue and spending**

Introduction

Climate change policy is evolving and **climate** finance is likely to be a crucial component of any future international climate **regime**. This chapter considers some of the key issues around climate finance and criteria for examining new proposals and funding schemes as they emerge, including those outside and within the **UNFCCC**. After examining the definition, rationale and requirements for climate finance, the chapter introduces some of the existing sources of international climate finance pertinent to developing countries. It then presents key analytical issues for climate finance arranged around three themes of governance, sources and spending. To date, most attention has been given to governance and sources, with far less emphasis on where and how to spend finance in ways that are simultaneously effective, efficient and equitable. Climate finance debates are situated within the broader context of a possible shift away from finance from development cooperation (aid) towards new forms of 'climate cooperation', with a different set of sources, governance and spending.

1. Finance for tackling climate change: What, why and how much?

Despite the significant attention to the subject in recent years, there is no agreed definition of what constitutes 'climate finance'. At the broadest level, climate finance can be interpreted to include both international public finance (bilateral aid, other official flows, export credits, and multi-lateral concessional and non-concessional flows) and private finance (carbon market finance, REDD+, Foreign Direct

Investment and other private flows) (Buchner *et al.*, 2011). Its interpretation within the context of the UNFCCC regime has stressed that finance must be new and additional to existing international and private financial flows, including ODA. However, it is unclear how **additionality** might be defined (see section 3.2 of this chapter).

1.1 What is the case for climate finance?

In terms of **social justice** and **equity**, the moral case for climate finance is based on the uneven distribution of causes and effects of climate change; the problem is caused primarily by richer people around the world and the costs will cause greater harm to poorer people. This rationale underpinned the original division of countries into the Annex groups with different commitments under the UNFCCC; higher emitting countries should act first in taking climate change actions and should also provide finance and access to technology for lower emitting (largely developing) countries (Shue, 1999).

Alongside these moral arguments, there are also instrumental imperatives for climate finance. For **mitigation**, financial costs of hitting any given **stabilization** target will be significantly higher if emissions reductions options in developing countries are not pursued. As these amount to 65–70 per cent of the total global emissions reductions potential, postponing mitigation in developing countries to 2020 could more than double the cost of stabilizing at a 2°C target for global warming (World Bank, 2010a). Another instrumental imperative for climate finance is based on limiting negative and enhancing positive impacts of climate change on development outcomes, including economic growth, well-being and poverty reduction (as highlighted in Chapter 2). If emissions continue unchecked, climate change impacts could lead to damages costing at least 5 per cent and possibly more than 20 per cent of global GDP, with impacts more severe in developing countries (Stern, 2007).

1.2 How much climate finance assistance is needed?

The level of required climate finance will depend on the target, timing and approach chosen:

- Costs will depend on the target chosen; mitigation costs will be higher for more stringent emissions reductions targets, and the cost of reducing each unit of emissions increases as the target becomes more stringent (as the low-cost actions are taken first). **Adaptation** costs will also grow the greater the extent of climate change and so are linked directly to the level of mitigation realized.
- The cost of achieving any given target for climate stabilization will be higher if action is taken later rather than earlier, prompting calls for early action to tackle climate change (Stern, 2006; UNDP, 2007).
- Costs will be higher if we do not pursue least-cost options, especially in emissions reductions (see explanations of the Marginal Abatement Cost Curve (MACC) in Chapter 4). Given that 70 per cent of the most economically feasible emissions reductions opportunities are in developing countries (McKinsey and Co., 2009a), an inclusive approach to mitigation is vital to minimize the costs of hitting a stabilization target. This rationale underpinned the Clean Development Mechanism under the Kyoto Protocol and remains an important instrumental factor driving climate finance to assist mitigation in developing countries.

There have been a wide range of analyses to estimate the required levels of climate finance (see Tables 6.1 and 6.2). While they are often crude approximations, these have played an important role in signalling the level and magnitude of international cooperation required to meet the challenge. While these numbers are large, they are small when compared to total global GDP (0.3–0.5 per cent) and as a proportion of total global investment (1.1–1.7 per cent) (UNFCCC, 2007b).

Four categories of mitigation measures are covered in these estimates (Olbrisch *et al.*, 2011):

1   Energy efficiency (buildings, industry and transportation).
2   Low-carbon energy supply (biofuels and renewables, nuclear) and CO_2 capture and storage (CCS) for electricity supply.
3   Reduction of other **GHG** emissions, including CCS for industrial emissions.
4   Carbon sinks (agriculture and forestry).

Table 6.1 *Estimated climate finance required for migration in developing countries**

Source of estimate	2010–20	2010–2030
Mitigation costs		
McKinsey & Company		175
Pacific Northwest National Laboratory (PNNL)		139
Mitigation financing needs		
International Institute for Applied Systems Analysis (IIASA)	63–165	264
International Energy Agency (IEA) Energy Technology Perspectives	565 (average to 2050)	
McKinsey & Company	300	563
Potsdam Institute for Climate Impact Research (PIK)		384

Source: World Bank, 2010a.

*US$ billions at 2005 prices, all estimates for stabilization of greenhouse gases at 450 ppm CO_2e.

Table 6.2 *Estimated climate finance required for adaptation in developing countries**

Review	Adaptation costs 2010–15	Adaptation costs 2030	Included measures
World Bank	9–41		Cost of climate-proofing development assistance, foreign and domestic investment
Stern Review	4–37		Cost of climate-proofing development assistance, foreign and domestic investment
United Nations Development Programme	83–105		Same as World Bank, plus cost of adapting Poverty Reduction Strategy Papers and strengthening disaster response
Oxfam	>50		Same as World Bank plus cost of National Adaptation Plan of Action and nongovernmental organization projects
UNFCCC		28–67	2030 cost in agriculture, forestry, water, health, coastal protection, and infrastructure
Project Catalyst		15–37	2030 cost for capacity building, research, disaster management and the UNFCCC sectors (most vulnerable countries and public sector only)
World Bank (EACC)		75–100	Average annual adaptation costs from 2010 to 2050 in the agriculture, forestry, fisheries, infrastructure, water resource management, and coastal zone sectors, including impacts on health, ecosystem services, and the effects of extreme-weather events

Source: World Bank, 2010a.

* US$ billions at 2005 prices, all estimates for stabilization of greenhouse gases at 450 ppm CO_2e.

Many of the initial investment costs required for energy efficiency measures can be recovered through lower energy bills, whereas the higher investment costs of low-carbon generation technologies are more likely to lead to higher electricity prices. Low investment costs are associated with enhancing carbon sinks, but they may have ongoing operating costs (Olbrisch *et al.*, 2011). Notably, many studies do not address the last two categories, reflecting their general under-emphasis in mitigation policy to date (Pielke, 2010).

Similar to mitigation estimates, studies of adaptation needs in developing countries have provided widely different results (see Table 6.2). Assessments vary depending on their specific methodologies, the projected amount of climate change assumed and the year for which their estimates are made. However, estimates have consistently demonstrated that the order of magnitude of the effort required is in the billions rather than millions of US dollars and there remains a significant gap between existing sources of international finance and estimates of what is required.

Many early adaptation cost assessments were based on calculating the global cost mark-ups for different investment flows based on how sensitive different sectors are to climatic factors (Agrawala, 2005; World Bank, 2006; Stern, 2007; UNDP, 2007). Later studies went further by breaking these down into different types of adaptation costs and examining the costs of both planned and private adaptation measures (UNFCCC, 2007a; World Bank, 2010b). Recognizing a bottom-up approach, others added the extra cost of NGO work at the community level and funding projects indicated by LDCs in their National Adaptation Programmes of Action (NAPAs) (Oxfam, 2007).

Guided by the UNFCCC definition of climate change (see Chapter 3), adaptation estimates have focused primarily on calculating the additional cost due to anthropogenic climate change, rather than the existing **adaptation deficit** for adaptation regardless of additional climate change as noted in Chapter 5. Fankhauser's (2010) review of adaptation costing found that many studies make assumptions about the additional 'mark-up' for adaptation on the basis of little empirical evidence, while Parry *et al.* (2009) have argued that the UNFCCC underestimates adaptation costs by ignoring some sectors, ignoring the pre-existing adaptation deficit and ignoring the residual loss and damage from impacts that will remain even after adaptation.

Estimates have also been critiqued for focusing too much attention on infrastructure changes and ignoring adaptation in terms of 'behaviour, innovation, operational practices, or locations of economic activity' (World Bank, 2010a: 260). This is because it is easier to calculate the costs of, for example, adding additional height to a sea defense than for soft measures such as early warning systems or community preparedness programmes, which depend on effective institutions supported by collective action (Narain *et al.*, 2011). Notwithstanding these methodological difficulties, estimates suggest significant additional costs of adaptation for development contexts such as achievement of the Millennium Development Goals (MDGs) (see Box 6.1).

Box 6.1

Climate-proofing the Millennium Development Goals in Africa

Fankhauser and Schmidt-Traub (2011) conducted an estimation of adaptation costs for the period 2010–2020 using an integrated approach to adaptation and development rather than treating adaptation as a stand-alone activity. The assessment used cost estimates of development interventions to meet the MDGs developed by the MDGs Africa Steering Group (Ban *et al.*, 2008) combined with sectoral estimates of adaptation costs (UNFCCC, 2007a).

The extra adaptation costs due to climate change arise from having to provide:

- more development support (for example, extra bed nets against malaria);
- the same support at a higher cost (for example, more expensive infrastructure);
- altogether new measures (for example, building institutional or adaptive capacity);
- a re-prioritization of certain measures compared to the baseline development plan (for example, enhanced disaster management activities due to an increasing hazard burden).

The study estimated that climate-resilient development in Africa could require international financial assistance in the order of US$100 billion a year over the period 2010–2020. This includes some US$82 billion in 'baseline' official development assistance and US$11–21 billion for incremental investments in adaptation. The total is about 40 per cent higher than the original MDG estimate of US$72 billion. This is higher than the incremental funding promised under the Cancun Agreements but, depending on how funds are allocated, not dramatically so.

Source: Fankhauser and Schmidt-Traub, 2011.

2. Sources and channels of climate finance

Climate change finance to developing countries runs through a range of different channels (see Figure 6.1), although sources are often mixed in any given delivery mechanism. This section discusses issues in relation to five groups of climate finance:

- UNFCCC mechanisms
- Development cooperation, bi-lateral and multi-lateral finance institutions
- Private sector finance
- National responses: Domestic finance and coordination
- Innovative funding proposals.

Figure 6.1 *Financial flows for climate change mitigation and adaptation in developing countries.*

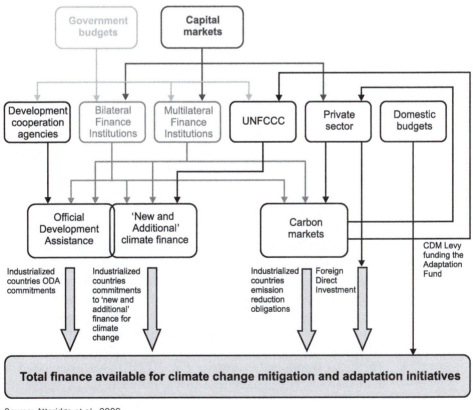

Source: Atteridge *et al.*, 2009.

Table 6.3 provides a summary of major climate finance instruments to date. However, this should be read in the context of the rapidly evolving policy environment, and readers should refer to the web-based platforms monitoring and tracking climate finance (see websites at end of chapter).

2.1 UNFCCC climate finance mechanisms

As summarized in Chapter 3, under the UNFCCC agreements, a number of international mechanisms have been established to assist developing countries in fulfilling their commitments to tackling climate change, as well as to provide flexibility to the richer countries in meeting their emissions reductions commitments. Two other major mechanisms, REDD+ and the Green Climate Fund, are under development (see Chapter 3 for more details on these).

Under the Kyoto Protocol, the Clean Development Mechanism (CDM) provides a means through which developing countries are able to receive financial assistance for taking actions to reduce **greenhouse gases**. Through the CDM, projects that generate emissions reductions receive credits (known as 'certified emissions reductions' or CERs), which can then be purchased by industrialized countries and contribute in part towards their domestic emissions reductions targets under the Protocol. Although it has received criticism for focusing on richer developing countries and providing little contribution to social benefits, it has played an important role as a major international market mechanism to simultaneously foster private sector investment, emissions reductions, and sustainable development in developing countries.

The other major funds under the UNFCCC are:

- The **Special Climate Change Fund** (SCCF), which finances projects in developing countries relating to adaptation, technology transfer and capacity building in a wide range of sectors including energy, transport, industry, agriculture, forestry and waste management, as well as economic diversification. In reality, this Fund has predominantly financed adaptation and forestry-related activities.
- The **Least Developed Countries Fund** (LDCF) was established to finance a work programme in LDCs, including the development

Table 6.3 Instruments and examples of climate funds

Type of instrument	Mitigation	Adaptation	Research, development, and diffusion
Market-based mechanisms to lower the costs of climate action and create incentives	Emissions trading (CDM, JI, voluntary), tradable renewable energy certificates, debt instruments (bonds)	Insurance (pools, indexes, weather derivatives, catastrophe bonds), payment for ecosystem services, debt instruments (bonds)	
Grant resources and concessional finance (levies and contributions including official development assistance and philanthropy) to pilot new tools, scale up and catalyze action, and act as seed money to leverage the private sector	GCF, GEF, SCCF, CTF, UN-REDD, FIP, FCPF, CDKN	GCF, Adaptation Fund, GEF, LDCF, SCCF, PPCR, CDKN	GEF, GEF/IFC Earth Fund, GEEREF, CDKN
Other instruments	Fiscal incentives (tax benefits on investments, subsidized loans, targeted tax or subsidies, export credits), norms and standards (including labels), inducement prizes and advanced market commitments, and trade and technology agreements		

Source: Authors update based on World Bank, 2010a and www.climatefundsupdate.org

Notes: CDM = Clean Development Mechanism; CDKN = Climate and Development Knowledge Network; CTF = Clean Technology Fund; FCPF = Forest Carbon Partnership Facility; FIP = Forest Investment Program; GEEREF = Global Energy Efficiency and Renewable Energy Fund (European Union); GEF = Global Environment Facility; IFC = International Finance Corporation; JI = Joint Implementation; LDCF = Least Developed Country Fund (UNFCCC/GEF); PPCR = Pilot Program for Climate Resilience; SCCF = Strategic Climate Change Fund (UNFCCC/GEF); UN-REDD = UN Collaborative Program on Reduced Emissions from Deforestation and forest Degradation.

and implementation of National Adaptation Programmes of Action.

- The **Kyoto Protocol Adaptation Fund** (known as the 'Adaptation Fund') to finance adaptation actions in developing countries. Financed from a levy of 2 per cent on the trading of CDM credits.

Both the SCCF and LDCF are financed from voluntary donations by Annex 1 countries and managed by the Washington-based Global Environment Facility (GEF, see Box 6.2)), which also manages the funding of Convention reporting requirements such as National Communications and Technology Needs Assessments. The GEF has formed an important source of funding for actions with international environmental benefits since its establishment just prior to the Rio Earth Summit in 1992. GEF financing mechanisms and procedures have provided a template that has helped shape climate finance. These include national ownership and disbursement to national governments via the assistance of international implementing bodies

Plate 6.1 *A Korean group promotes filling the Green Climate Fund's piggy bank at COP 17 in Durban.*

Credit: WRI /Michael Oko.

Box 6.2

The Global Environment Facility

The GEF was established in 1991 to provide funding to developing countries to protect the global environment. It covers thematic areas that include biodiversity, international waters, ozone-depleting substances, land degradation, and persistent organic pollutants, and climate change. By the end of its fourth funding cycle it had disbursed over US$8.7 billion in project finance to developing countries, in addition to acting as the financial mechanism for the UNFCCC. It funds full costs for reporting activities under the UNFCCC such as National Communications, but the financing of other climate change actions in developing countries is based on financing the additional, 'incremental' costs of actions (as it is for its wider, non-UNFCCC funding).

This principle of incremental cost has proved controversial in part due to the difficulties in separating climate change from broader development actions, and separating anthropogenic climate change impacts from background climate variability. The GEF has also faced criticism by developing countries over the complicated procedures for accessing finance and for governance structure dominated by the richer donor countries (Gupta, 1997). Its suitability as a manager of adaptation finance has also been questioned (Möhner and Klein, 2007), while others have argued that the GEF's location in the World Bank structure is representative of a neoliberal development ideology that consistently gives insufficient attention to environment issues (Young, 2002).

such as UNDP, UNEP, FAO and the World Bank, as well as the principle of calculating and funding the incremental cost of actions. Incremental cost principles determine that international finance is disbursed only for the extra adjustment needed due to climate change rather than the baseline costs of action that is required regardless of climate change.

Although open to other sources of finance, the Adaptation Fund is financed primarily from the market for CDM carbon credits rather than voluntary donations. This is an important distinction as it should provide a more stable source of revenue, although this depends on an active market for CDM emissions reductions credits. However, the tax-based source of revenue and the Fund's management by a Board dominated by developing country representatives provides a different approach to conventional aid financing (see Box 6.3).

Recognizing that these existing mechanisms were not providing levels of finance commensurate with the climate change challenge, the UNFCCC decisions at Cancun in 2010 agreed to develop a Green

Box 6.3

The Adaptation Fund

The Adaptation Fund (AF), in contrast to the other Convention funds, is financed primarily from a levy of 2 per cent on trading of CDM credits. Although revenues vary depending on CDM market trading, it is regarded as a more predictable source than other UNFCCC funds, which rely on voluntary donations from richer countries. Rather than using the GEF as a financial mechanism, the AF is managed by a Board comprised of members appointed by and accountable to the UNFCCC Kyoto Protocol governing body (CMP). It also uses a direct access model whereby accredited national institutions are able to apply for and manage grants rather than having to work with an international intermediary partner as with the GEF-managed sources. It also changes the rationale for adaptation finance from a charitable basis to one of compensation for the damages caused by climate change impacts, and alters power relationships through a governance structure that reflects equity in decision-making among developed and developing countries over how money should be distributed and spent (Ayers and Huq, 2009). The status of the AF with regard to the development of the Green Climate Fund is unclear, although there are calls for it to maintain a role as the adaptation funding mechanism within the new Fund (Müller and Khan, 2011).

Despite its ground-breaking nature, the AF has initially received relatively small amounts of finance due to a range of factors (Harmeling and Kaloga, 2011: p. 25):

● At low carbon prices the CDM levy alone does not generate significant resources. Some developing countries have resisted additional voluntary contributions as they affect the Fund's independence and self-sufficiency.

● Donors were hesitant to support the AF initially due to its innovative features, uncertainty over the effectiveness of the Board and national implementing agencies, and the Fund's focus on separate projects and programmes rather than climate proofing of investments and mainstreaming of adaptation.

● ODA is usually channelled through developing countries' finance ministries, whereas the UNFCCC processes are usually in the hands of environment ministries, and there is hesitance on the donor side to deal with the unfamiliar pathways on a significant financial scale.

Climate Fund (GCF) in order to scale up the provision of long-term financing for developing countries. The agreements also called for Annex 1 countries to mobilize new and additional 'fast start' finance of roughly US$30 billion during the period required for establishing the Green Climate Fund. Design of the governance, rules and procedures of the GCF is coordinated by a Transitional Committee made up of UNFCCC representatives.

The Cancun texts primarily agreed overall governance of the GCF, including accountability to the UNFCCC Conference of the Parties body, governance by a Board of 24 members composed of an even balance between developed and developing countries and between regions, including representation for particularly vulnerable countries (SIDS and LDCs). The GCF faces a range of challenges not just in its governance, but in how it will mobilize funding, how to allocate between countries and between different priorities of adaptation, mitigation and REDD+/agriculture, and what mechanisms to use to disburse the funding (Bird *et al.*, 2011a). Final agreement on the Fund is a significant pillar of the post-2020 agreements due for completion in 2015 under the Durban Platform for Action. The GCF is discussed further in section 2.5 in this chapter.

2.2 Bilateral and multi-lateral climate finance

Climate change has now moved from being a marginal issue related to energy and environmental management to become a core issue in international development cooperation (G8, 2005; World Bank, 2010a). There is increased acknowledgement among development cooperation agencies and their partners that climate change impacts can create investment risks that threaten the achievement of their objectives, including the Millennium Development Goals (MDGs).

Development investments may also affect the **vulnerability** of societies to climate change impacts and their ability to respond. According to estimates by the World Bank, up to 40 per cent of ODA-financed development activities are sensitive to climate risk (World Bank, 2006). Estimates by the OECD suggest that more than 60 per cent of all ODA could be relevant to building **adaptive capacity** and facilitating adaptation (Levina, 2007). At the same time, mitigation and adaptation actions may provide synergistic opportunities to enhance progress towards development objectives (AfDB *et al.*, 2003; UNDP, 2007).

As a consequence of these risks and opportunities, there have been two simultaneous responses from bilateral and multi-lateral development agencies and their developing country partners. The first has been an inward-looking effort to integrate, or mainstream, climate change considerations into development co-operation activities, sometimes termed 'climate proofing' (Hammill and Tanner, 2011).

This **mainstreaming** challenge has been met in particular through the development of a burgeoning variety of tools and learning resources to improve awareness and decision-making (Hammill and Tanner, 2011; Tanner *et al.*, 2012; www.climateplanning.org). For adaptation, mainstreaming includes tackling the risks of damages from climate **hazards** to specific programme investments, the risk of general under-performance of the investment due to climate change impacts, and attempting to avoid **maladaptation** that inadvertently increases vulnerability to climate change (van Aalst, 2006; Barnett and O'Neil, 2010). From the low carbon perspective, donors have focused especially on 'development first' and 'win-win' investment opportunities that simultaneously provide development as well as emissions reductions benefits (Bradley and Baumert, 2005; Kok *et al.*, 2008). This is in part due to the political sensitivity of imposing emissions reductions conditions on aid disbursement (Kok *et al.*, 2008).

At the same time as this mainstreaming drive, there has also been an expansion in the number and size of specific climate change financing mechanisms, many of which are independent from the UNFCCC mechanisms. These include bilateral and multi-lateral programmes such as the Climate and Development Knowledge Network (CDKN) and the World Bank-led Climate Investment Funds (CIFs) (see Table 6.3 and Box 6.4). The CIFs were created as a temporary measure to bridge the financing and learning gap prior to agreements on post-Kyoto financing. They have attracted far greater financial contributions from donors than the UNFCCC funds and their evolving relationship with the Green Climate Fund will have a bearing on the future climate finance architecture (see Chapter 3).

2.3 National responses to climate finance

Climate finance has also begun to emerge as a major development issue at the national level. The NAPA and NAMA reporting processes under the UNFCCC (see Chapter 3) communicate national adaptation and mitigation needs to international climate funding mechanisms. The National Adaptation Programmes of Action (NAPAs) were designed to communicate urgent and immediate adaptation needs with a view to accessing finance for implementation of interventions that could tackle these needs. Total estimated funding required to implement the priority NAPA adaptation projects in the Least Developed Countries is US$1.66 billion (LDC Expert Group, 2009).

Box 6.4

The Climate Investment Funds (CIFs)

The CIFs are a set of World Bank-managed Funds that aim to provide concessional loans and grants for policy reforms and investments that achieve development goals through a transition to a low carbon development path and a climate-resilient economy (World Bank, 2008). The CIFs are comprised of two Trust Funds:

(1) The **Clean Technology Fund (CTF)** promotes investments to initiate a shift towards clean technologies. The CTF provides concessional rates of finance at a scale necessary to incentivize developing countries to integrate nationally appropriate mitigation actions into sustainable development plans and investment decisions.

(2) The **Strategic Climate Fund (SCF)** serves as an overarching fund to support programmes aimed at a specific climate change challenge or sectoral response, including:

- The Forest Investment Program (FIP) supports developing countries' efforts to reduce emissions from deforestation and forest degradation by providing scaled-up bridge financing for readiness reforms and public and private investments. It finances programmatic efforts to address the underlying causes of deforestation and forest degradation and to overcome barriers that have hindered past efforts to do so.

- The Pilot Program for Climate Resilience (PPCR) aims to pilot and demonstrate ways in which climate risk and resilience may be integrated into core development planning and implementation. In this way, the PPCR provides incentives for scaled-up action and initiates transformational change.

- The Program for Scaling Up Renewable Energy in Low Income Countries (SREP) aims to demonstrate the economic, social and environmental viability of low carbon development pathways in the energy sector by creating new economic opportunities and increasing energy access through the use of renewable energy.

Source: www.climateinvestmentfunds.org

Whilst much attention has focused on international climate finance, there has also been a growth in national revenue-raising measures through tax reform (including introduction of carbon taxes) and market development. Countries including China, Costa Rica, India, Indonesia and South Africa have all established or tabled domestic carbon taxes, often applied to specific sectors or products such as high intensity manufacturing or fossil fuel consumption. As well as contributing to global emissions reductions, these efforts commonly have the aim of domestic social and economic benefits such as promoting energy efficiency to enhance profitability, renewable energy development for energy security, reducing air pollution, subsidizing affordable low-energy appliances for poor households, or funding payments for ecosystem services.

Integration of international flows of climate finance with such public sector flows is important for ensuring fiscal stability in developing countries, reducing the risk that flows of climate finance distort the domestic economy and public sector (World Bank, 2010a). Multiple and uncoordinated sources of climate finance may add to the administrative and institutional requirements and potential for fragmentation of national systems (Atteridge *et al.*, 2009). As a result, there are growing examples of national efforts to coordinate climate finance, including both international and domestic sources of finance.

One response to the fragmentation of different sources of climate finance has been the creation of country-level, multi-donor trust funds. These are intended to coordinate donor and government decision-making, link existing national development and climate change policy frameworks to produce greater coherence, reduce reporting burdens, and increase transparency and accountability (Lockwood and Cameron, 2011). Examples are now widespread, including in Indonesia, Bangladesh and Ethiopia (WRI, 2011; Alam *et al.*, 2011; Lockwood and Cameron, 2011). The Indonesia Climate Change Trust Fund (ICCTF) was established as an independent entity operated by the Ministries of Planning (BAPPENAS) and Finance, designed to address the immediate and emerging needs of Indonesia's Climate Change Sectoral Roadmap (CCSR) programme investments. It is focused on moving the economy to low greenhouse gas emissions, to improve climate and improve leadership and management for tackling climate change. The ITCF explicitly attempts to integrate international finance with national investment policies and to facilitate private sector investment in climate change.

By contrast, the Bangladesh Climate Change Resilience Fund (BCCRF) is more explicitly focused on adaptation concerns, reflecting national priorities. It aims to blend international finance with domestic revenue allocations and to support the national Climate Change Strategy and Action Plan. The role of the World Bank in the Fund's governance has been a source of controversy, whilst there are some features of the BCCRF that may not work well for **low-carbon development** finance: its heavy adaptation focus, little engagement with the private sector, and difficulty of financing relatively small projects, such as small scale renewable in the early phase of scale-up.

2.4 Climate finance and the private sector

Given the scale of estimated financing needs to tackle climate change in developing countries, public sources alone will not be able to meet these needs. Although governments play a major role in infrastructure investments that affect the opportunities for low carbon production, they account for less than 15 per cent of global economy-wide investment (World Bank, 2010a). Strengthening the role of the private sector in tackling adaptation and mitigation in developing countries is therefore vital (AGF, 2010). This mix of public and private sources is acknowledged by the UNFCCC agreements, although clarity on their respective roles, mechanisms, and accounting systems remains limited. The private sector can be seen either as an upstream source of finance, as a downstream recipient of the multi-lateral financing institutions such as the World Bank, or in parallel as a completely separate and independent finance stream (Atteridge *et al.*, 2009; see Figure 6.1 earlier).

Much of the attention regarding low-carbon development issues has focused on private sector participation in carbon markets. Despite the global economic downturn and collapse in carbon price prompted by oversupply of carbon credits in the European Union Emissions Trading Scheme (EU ETS) and the weakness of the second Kyoto commitment period, the value of the global carbon market continues to rise, reaching US$176 billion by 2011 (World Bank, 2012a). This is remarkable given that the market and related carbon reduction activity was almost non-existent in the early 1990s. The carbon market includes activity in the Clean Development Mechanism (CDM), through which emissions reductions activities in developing countries generate carbon credits that can be traded on the carbon market (see Chapter 3). Voluntary carbon offset schemes and nascent REDD+ initiatives have also provided opportunities for private sector participation. From a development perspective, however, low demand for credits from industrialized countries has meant that transfers, while significant, have been 'modest relative to the amount of mitigation money that will have to be raised' (World Bank, 2010a).

Providing the global private sector with signals of longer term stability (and growth) of the carbon price is a crucial component of the post-2020 deal due for completion in 2015. In the absence of a mature global carbon market, efforts have shifted towards how public sector actions can reduce the risks to private sector investment in order to

leverage private capital for mitigation, adaptation and development
(see Box 6.5). These risks include those generic to all investment in
developing countries, and those more specific to low carbon
investments, including unfamiliar risks in dealing with a new sector. A
variety of instruments are therefore now being used to address real or
perceived risks, either through direct public financing or through
public guarantees that cover risks through insurance-like tools (Brown
and Jacobs, 2011).

For adaptation, the role of private sector finance is complicated by the
absence of a clearly marketable commodity (Persson, 2011) and the
fact that many adaptation actions are public goods, such as flood
protection, early warning systems or providing climate information, so
often require governments to provide them (see Chapter 8). Recent
experiences suggest that insurance could be a critical mechanism to
bring together public and private sectors for adaptation. Insurance has
the potential to promote or create a virtuous cycle that improves
ability to access credit, encourages investment in productive assets

Box 6.5

Public climate financing instruments used to leverage private capital

The private sector seeks markets with both attractive returns relative to associated risks
and adequate size, liquidity, and transparency. Both of these conditions are often absent
in developing countries. Public sector financing can therefore be used to help manage
these risks and leverage private capital for climate change actions. These include:

For **existing political and macroeconomic risks**: instruments to manage risk may
include providing political risk guarantees, interest-rate/currency exchange
products, and local currency loans can help investors and project developers
financially manage political (for example, political instability) and/or
macroeconomic (for example, exchange rate volatility) risks.

For **low-carbon market risks, including policy, technology, and operational
risks**: In newer low-carbon markets, instruments to manage risk may include public
financing instruments like first-loss equity and debt investments and concessional
loans can be instrumental in encouraging early investment. Projects in more
established low-carbon markets – like solar, wind, and energy efficiency – can
benefit from flexible loans, partial risk and credit guarantees, and risk sharing
facilities.

Source: Adapted from Venugopal and Srivastava, 2012.

and higher risk/higher yield returns that can help people escape poverty, and incentivize risk reduction measures (Warner *et al.*, 2009). Linking insurance to the UNFCCC, Linnerooth-Bayer and Mechler (2006) set out a two-tiered strategy. First, an insurance programme to help develop insurance-related instruments that are affordable to poor people and linked to incentives for proactive adaptation measures. Second, a complementary disaster programme to assist with disaster relief, provided countries can demonstrate they are making efforts to reduce and manage risks.

There has been particular interest in crop insurance to support developing country agricultural sectors. Conventional indemnity-based crop insurance, where claims are based on crop losses reported by claim adjusters, are increasingly being replaced by weather-indexed crop insurance schemes. These schemes have attempted to overcome the frequent failure of traditional insurance to extend cover to the poorest and most vulnerable people and regions. They create a contract written against a weather index, usually based on an established relationship between weather events and crop failure. Rather than having to have their losses assessed, farmers collect immediate insurance compensation if the index reaches a certain weather-related trigger (such as a number of continuous dry days), regardless of actual losses. Although not without their challenges (see Box 6.6), weather-indexed schemes attempt to break cycles of impoverishment following weather-related shocks and overcome the problems of traditional indemnity-based insurance that include high costs of verifying losses, 'moral hazard' that reduces risk taking, and adverse selection of crops based on an expectation of payouts for poor harvests (Hess and Syroka, 2005; Hellmuth *et al.*, 2007).

Box 6.6

Challenges of weather-index based insurance schemes

Assessing the success of weather-indexed insurance schemes remains difficult as examples are in pilot stage and none have experienced a major and widespread catastrophic event. The multiple pilot schemes around the world have faced significant challenges: 'covariate' climate risks affecting the majority of populations make risk spreading difficult; marginal and subsistence farmers remain difficult to target; an increasing hazard burden may affect financial sustainability; and there are significant capital costs for start up and operation (Mechler *et al.*, 2006; Hochrainer *et al.*, 2007; Meze-Hausken *et al.*, 2009).

Most studies and pilots suggest the need for blending private sector finance with some form of public subsidy either from donors or national governments. In Mongolia, the Index-Based Livestock Insurance Project helps herders cope with significant herd losses and transfer some of the risks of raising livestock, relying on external assistance to create high-quality data on livestock mortality, outreach efforts to educate herders on the insurance products, and capacity building by government officials and insurance companies (Mahul and Skees, 2007; Luxbacher and Goodland, 2010). Evidence from the packaged loan and index-based insurance products in Malawi suggests potential for smoothing disaster shocks among low-income and low-asset households (Hochrainer *et al.*, 2007; Suarez and Linnerooth-Bayer, 2010). However, this and other schemes would not be possible without external assistance due to affordability, lack of an insurance tradition and market, start-up and monitoring costs and technical assistance (Hochrainer *et al.*, 2007).

Another frequent constraint is the affordability of insurance premiums. The Horn of Africa Risk Transfers for Adaptation (HARITA) programme piloted in Ethiopia combines community identified climate resilience projects (risk reduction), micro-insurance (risk transfer), microcredit ('prudent' risk taking), and savings (risk reserves). Crucially, it enables poorer farmers to exchange their labour on community risk reduction projects for an insurance certificate protecting them against deficit rainfall. While this approach also requires external financing, it aims to deliver value for money by simultaneously providing insurance and risk reduction actions (WFP and IFAD, 2010).

A final question arises for payment of premiums for crop or other damage arising from climate change: is it ethical to expect poor people to pay such premiums, given that the risks they face are not of their making? This question refers to the potential role of the main carbon emitters to contribute, with adaptation funding as a source of insurance underwriting or premium subsidy.

2.5 Proposals for future climate finance

The gap between the estimated requirements for climate finance and the current investment flows has led to a range of proposals and debates around new and innovative climate finance mechanisms (Müller, 2008; Bapna and McGray, 2008; Brown, 2009; AGF, 2010; Hof *et al.*, 2011). The Cancun agreements of 2010 made a major step in setting targets for climate finance. However, given the scale of estimated needs, the pressure to reduce government spending in many countries, and the need to raise finance in addition to existing aid commitments, other sources will need to be generated (World Bank, 2010a). While private sector finance has made some progress in low carbon markets in developing countries, it is less clear how these investment flows can explicitly benefit poor people, particularly for adaptation where the profit incentives may be weak and the risks high. As a result, a range of different proposals for innovative climate finance have been put forward (UNFCCC, 2008).

Some proposals suggest generating additional finance from carbon markets. One approach is to place a levy (tax) on carbon trading, following the model adopted by the Adaptation Fund, and there have been proposals to increase this 'share of the proceeds' from the CDM from the current 2 per cent or extend the levy to other instruments such as Joint Implementation or the EU ETS. This model has been critiqued for providing perverse incentives as it taxes emissions reductions (a good thing), rather than the emissions themselves. If it restricts emissions trading, it may also reduce benefits to developing countries (Müller, 2008).

Another carbon market approach seeks to set aside a fraction of the allocation of each country's emissions allowances agreed under an international emissions reductions agreement. Adaptation finance would then be generated by auctioning these allowances to the highest bidder (including bidders from the private sector). Originally proposed by Norway for the Kyoto Protocol, this could operate under a post-2020 agreement and would crucially generate finance outside national budgets whilst avoiding the perverse incentives of a levy-based approach (Hof et al., 2011). Such an approach could also be applied to auctions of allowances under national carbon markets as they emerge, or to the existing EU ETS.

An international tax on carbon would also avoid perverse incentives by taxing emissions themselves rather than emissions reductions. An international approach would build on the emergence of national carbon-related taxes in many countries to provide an international financing component. Although an international carbon tax would require a complex international and national regulatory mechanism, it would also have the advantage of a broad and secure base for climate finance (World Bank, 2010a). The Swiss government has proposed a global carbon tax of US$2 per tonne of CO_2 on energy-related CO_2 emissions. It included a basic exemption for each country of 1.5 tCO_2 per capita to take into account the principle of common but differentiated responsibilities and **carbon space** for development, designed to benefit the countries with the lowest emissions per capita (UVEK, 2008). However, an international tax would impinge on the tax-setting power of sovereign governments and is therefore unlikely to gain unanimous agreement. Nevertheless, carbon-based taxation is likely to become more international, even where it is not framed explicitly on carbon, for example in China's tax on high energy intensive exports.

Other proposals have focused on specific taxation of international aviation or shipping. These sectors are currently not included in emissions reductions commitments and are frequently exempt from national duties, despite rapidly growing emissions trajectories. As such, these proposals could be seen as closing current loopholes rather than applying new taxation. For example, the Tuvalu Adaptation Blueprint proposes a levy on international airfares and maritime transport freight charges, the International Air Travel Adaptation Levy (IATAL) proposes a levy on air travel emissions or ticket price, while the International Maritime Emission Reduction Scheme (IMERS) proposes a levy on international maritime transport emissions at the same level of average market carbon price (UNFCCC, 2008). However, the suggested levies are too low to generate significant funding, while the administrative hurdles are considerable given the power of international bodies such as the International Maritime Organization (Müller, 2008; World Bank, 2010a).

3. Assessing climate finance: Governance, sources and disbursement

Climate finance is evolving rapidly. This section provides a range of analytical issues to assess options as they emerge, based on the division of finance into governance, sourcing and disbursement.

3.1 Governance of climate finance

Governance is crucial to channelling financial flows to respond to climate change, but there are a wide range of disparate sources of finance, and generally poor accountability to those most affected by climate change. 'Efforts to strengthen the architecture of climate governance will therefore have to build in safeguards against risk, including corruption risks, in order for decisions made to have collective ownership, legitimacy and, ultimately, meaningful effect at the international, national and local levels' (Transparency International 2010:4).

Much of the debate in operationalizing funds has been around what institutional mechanisms can best govern this finance and its delivery, both at international and national level. Internationally, there has been competition over the control of climate finance between the institutions of the United Nations and those of the Bretton Woods system, which include the World Bank and the GEF. Schalatek *et al.* (2010: 3) note how at the Copenhagen Summit 'the World Bank and MDBs were heavily represented in a massive sales pitch to persuade the ministers and heads of states present to channel the promised fast track financing for adaptation and mitigation through the Climate Investment Funds (CIFs) (particularly, since some of the funding included in the figure of US$30 billion over three years might have already been previously committed to them)'.

The World Bank's development of the CIFs is regarded by some as paving the way for 'a longer-term shift in climate finance sources and delivery mechanisms which establish a longer-term role for the World Bank and the MDBs in climate finance and implementation' (Seballos and Kreft, 2011: 39). Despite this, however, the CIF funds contain 'sunset clauses', meaning their lifetime is intended to serve as a bridge to the development of a fully functioning climate funding mechanism – in principle the Green Climate Fund, for which the World Bank is working as the trustee during its development period.

Although the simple bipolar 'North-South' division of countries is increasingly less clear cut (see Chapter 3), a rough divide remains; the industrialized countries of UNFCCC's Annex 1 generally favor use of the pre-existing ODA and multi-lateral development bank institutions for governance and delivery, while the non-Annex 1 developing countries seek to establish new institutions outside the ODA aid architecture. Creating a shared sense of legitimacy over international governance of climate finance remains a crucial building block of the future regime (Roberts and Parks, 2007). Ballesteros *et al.* (2010: viii) suggest that rather than to debate *which institutions* should govern new flows of climate finance, it is more instructive to examine how governments can design a climate financial mechanism in a way that is widely perceived as legitimate.

At national level and local, the governance of climate finance faces many of the same challenges of development finance, including the difficulties of managing multiple sources of finance as well as ensuring effective delivery. Reviews of climate financing suggest that there remain major challenges for national climate finance around budgeting, targeting, decentralization, focusing on results, transparency and the role of the private sector (Bird, 2011). Lockwood and Cameron (2011) summarize key issues from OECD reviews in five countries in the Asia-Pacific region and six African countries, including:

- In many countries, ownership of climate change as an issue is limited, and so financing is largely supply driven, rather than demand led. No country had a dedicated forum for dialogue on climate financing involving donors, government, civil society, and the private sector.
- In Africa, responses to climate change only become coherent when politically important ministries become engaged. In many circumstances, climate finance flows to Environment Ministries, which are often politically marginal.
- Mechanisms for donor coordination at country level existed but were not functioning fully. This is a significant risk given the increasing diversity of different sources of climate finance.
- Effective use of climate finance requires a combination of elements – overall vision and policy, action plans, targets and budgets, functioning institutional arrangements and a mechanism of accountability, that was found nowhere in Africa. This weak country capacity means it is hard to integrate climate finance into local budgets.
- Where capacity is stronger and fiduciary risk is lower, 'direct access' arrangements, such as routing finance through a national development bank, would offer more alignment with local budgets (similarly to direct budget support), but there is limited experience with this approach.

3.2 Sources of revenue: Raising climate finance

A range of different criteria has been used to examine different potential sources of international climate finance (see Box 6.7; AGF, 2010; Hof et al., 2011; Schalatek and Bird, 2011).

Equity and social justice imperatives underpin moral arguments that climate finance should not be treated in the same way as conventional development cooperation (ODA) (Action Aid, 2007; Oxfam International, 2007; Ayers and Huq, 2009; Müller and Khan, 2011). The implication is that the existing institutions and governance of development cooperation, the conditions placed on development cooperation flows, and the dominant relationship of the donors are all

Box 6.7

Criteria for assessing sources of climate finance

The High-level Advisory Group on Climate Change Financing (AGF), established by the UN Secretary General following the Copenhagen Summit to study the potential AGF, developed a set of criteria in order to assess potential sources of climate finance, including:

● **Efficiency**: How a given source contributes to creating a 'price' to correct for the carbon externality and the potential impact of the source on growth or other risk factors.

● **Equity**: The incidence, or 'who really pays', for the source of finance. To be equitable, the direct burden of raising revenues should fall predominantly on developed countries.

● **Practicality**: The feasibility of implementation. For example, in the required institutional design and in relation to rules and laws in different countries.

● **Reliability**: The extent to which the source of finance is likely to lead to a predictable revenue stream in the future.

● **Additionality**: The extent to which new resources add to the existing level of resources, especially against existing ODA. 'Newness of a source' can form a useful proxy for additionality.

● **Acceptability**: Sources must be politically acceptable to both developed and developing countries. This illustrates the importance of having a variety of financing instruments available.

Source: AGF, 2010.

inappropriate for climate finance and need re-thinking for 'climate cooperation'. Climate finance therefore requires:

> . . . a new global relationship, shaped by 'polluter-pays' rather than charity. It means: governance structures to allocate money weighted toward developing countries; resistance to the use of traditional aid rules and conditions; new lines of accountability to the United Nations Framework Convention on Climate Change (UNFCCC), rather than to traditional 'aid givers'; and demands from developing countries for direct access to finance, without the need to work through intermediaries.
>
> (Mitchell and Maxwell, 2010: 2)

Many developing countries therefore want to see international climate finance directed through the UNFCCC institutions rather than through the aid architecture. There has been particular controversy over the additionality of climate finance (Stadelmann *et al.*, 2010), 'the extent to which new resources add to the existing level of resources (instead of replacing any of them) and result in a greater aggregate level of resources' (AGF, 2010: 26).

Establishing such additionality is difficult because of how to set the baseline against which to determine anything additional, but also in determining what flows of finance to include. Where loans are made to fund climate projects on a concessional (reduced interest rate) basis, such as in the World Bank's Climate Investment Funds, it is unclear if climate finance should be considered as the entire loan, or only the grant element that reduces the interest rate on the loan (Seballos and Kreft, 2011). Similarly, where public sector finance has provided the risk reducing incentive or enabling environment for additional private sector investment, it is unclear whether to count the overall investment leveraged or only the cost of the enabling actions. For example, should only the US$3 billion allocated for the Climate Investment Funds (CIFs) in 2009 be included as climate finance, or should the reported sum of an additional US$27 billion in leveraged finance, some of it from the private sector, be added to the total?

Additionality has been particularly difficult to evaluate where climate finance is delivered through ODA because of the mainstreaming of climate change objectives into development policy, programmes and projects. With many donors struggling to meet their existing ODA commitments, there is concern that allocating climate change finance from ODA may displace finance from other areas such as health or education (Oxfam International, 2007; Action Aid, 2007; Stadelmann *et al.*, 2010). This has led to calls to ensure that climate finance is raised through mechanisms that can be accounted for and governed separately from ODA and other pre-existing financial flows (Ayers and Huq, 2009). One suggestion for additionality has therefore been to treat the newness of a source as a useful, if partial, proxy (AGF, 2010).

3.3 Disbursement and delivery: Spending climate finance

While significant attention has been paid to the sources and governance of climate finance, the issue of how it is spent remains under-explored. This includes how to determine its allocation to countries, sectors and activities and how it is accessed and implemented.

Allocation criteria to determine spending priorities for climate finance have proven problematic. Decisions on mitigation actions are not guided only by cost-efficiency criteria (as often assumed by economics models), but also by development co-benefits, vested interests, uncertainty and political considerations around international equity. Similarly, various attempts to identify transparent international prioritization criteria for adaptation finance according to vulnerability have yielded different results depending on scientific approach used (see Box 6.8). Evidence from the application of prioritization criteria in funding mechanisms such as the Global Climate Change Alliance

Box 6.8

How can we prioritize adaptation finance according to vulnerability?

The Convention commits developed countries 'to assist developing countries that are particularly vulnerable to the adverse effects of climate change in meeting costs of adaptation to those adverse effects', but does not define how to identify these 'particularly vulnerable' countries. Many studies and funds have attempted to do so, but are commonly determined by subjective, political decisions over which indicators to use. In reality, allocation of many funds is driven by which countries express the strongest demand.

Füssel (2010) presents a semi-quantitative analysis of the difference between countries' responsibility for climate change, their capability to act and assist, and their vulnerability to climate change. He reveals complex and geographically heterogeneous patterns of vulnerability, and argues that the allocation of international adaptation funds to developing countries should be guided by sector-specific or hazard-specific criteria.

Hinkel (2011) finds that vulnerability indicators are only appropriate for identifying vulnerable people, communities and regions, and then only at relatively large (i.e. local to national) scales, when systems can be narrowly defined and inductive arguments can be built. He concludes that for allocation of adaptation funds, either vulnerability is not the adequate concept or vulnerability indicators are not the adequate methodology. Indeed, 'measuring' vulnerability is impossible and raises false expectations.

Klein (2009) notes the many possible interpretations of what constitutes 'particularly vulnerable'. He argues that comparing the potential impacts of climate change on, for example, human life, physical infrastructure or biological diversity eventually requires a subjective judgement about which outcomes are more desirable. Science can contribute (e.g. by proposing methods and collecting data), but cannot develop a definitive, objective and unchallengeable method to rank countries according to their vulnerability to climate change.

Source: Adapted from Klein and Moehner, 2011.

(GCCA), the Pilot Program for Climate Resilience (PPCR) and the Adaptation Fund suggests that the criteria applied vary and other political criteria also play a role (Klein and Möhner, 2011).

At a broader level, there are trade-offs between investments in mitigation versus adaptation. This involves evaluating not only the local costs and benefits of actions taken but also the global benefits of mitigation (a global public good) over adaptation, which primarily benefits the location in which it is undertaken. In reality, the split between allocation for adaptation and mitigation is politically negotiated, informed particularly by developing countries' demands for adaptation assistance and the need for mitigation actions to be focused predominantly in the richer countries. In reality, the separation of funding mechanisms for adaptation, mitigation and development may prevent the development of 'triple win' actions that simultaneously contribute to multiple adaptation, mitigation and development goals.

There is also contention between the use of market and fund-based instruments, and of grants versus loans. A market-based approach, particularly to mitigation and carbon markets, is seen by some as the most efficient means of achieving climate change related goals, as well as providing the incentives for private sector engagement. Developing countries have generally pushed for the use of specific funds, however, and have generally supported grant-based finance over concessional loans, arguing that responsibility for assisting the developing countries to tackle climate change must be additional and that finance is not owed to poor countries as 'aid' but as compensation from high-emission countries, which have caused the problem (Ayers and Huq, 2009; Oxfam International, 2007; Action Aid, 2007).

Regarding access to the funds, developing countries have generally supported 'direct access' arrangements, such as that employed by the Adaptation Fund. This implies developing countries directly accessing international public financing to implement national and local actions. Under direct access, national bodies take on the facilitation and project management function that has previously been undertaken by multi-lateral, international, and bilateral entities (such as UNDP, UNEP, World Bank or bilateral aid programmes) (Bird *et al.*, 2011b). Building direct access options into the Green Climate Fund is a key demand of developing countries in the UNFCCC negotiations, but evidence from the Adaptation Fund experience suggests that significant capacity building support is needed to enable these national and sub-national implementing entities to meet required standards of financial management (Schalatek *et al.*, 2012).

Additionality concerns result in a frequent requirement for climate finance mechanisms to be explicit in separating the climate change component from funding for non-climate-related factors. This is despite the fundamental interconnectedness of most climate change and development responses (McGray *et al.*, 2007; Kok *et al.*, 2008). As a result, this may bias funding towards certain sectors or activities where such separation is easier to calculate, particularly those where new approaches are technology-based or relate to 'hard' systems such as proofing of infrastructure, rather than 'soft' systems such as changes to management or building broader adaptive capacity (Fankhauser and Burton, 2011).

Finally, there is widespread acknowledgement that climate finance requires more thorough measurement, reporting and verification (MRV) processes. MRV processes are commonly seen as crucial for tracking finance against commitments, but must also evolve to evaluate the results they achieve, if the money is spent effectively and to enable learning to improve future financing systems and resulting actions. This is made harder by the lack of agreed definition of climate finance and the blurred boundary between public and private finance noted in earlier sections. Current MRV systems for different financial mechanisms and flows also lack transparency, comprehensiveness and comparability. Buchner *et al.* (2011: p. 8) suggest four action items to support MRV as part of a future global climate deal:

- **Adopting clear definitions of climate finance** spanning both public and private sources, and standardizing tracking of international climate finance flows from both a donor and a recipient perspective.
- **Exploring various different approaches to tracking climate finance** within a more comprehensive MRV system, drawing the lessons from existing information systems.
- **Improving reporting of public climate finance flows** from both a donor and a recipient perspective, building on existing information systems, ongoing efforts to improve these (e.g. national communications) and new reporting tools established under the Cancun Agreements (i.e. biennial reports, registries).
- **Extending reporting to include basic reporting of private climate finance**. A minimum level of information could be

ensured by requesting public finance sources to report on leveraging ratios and by streamlining the reporting on finance flowing through carbon markets.

Conclusion

By way of summary, Table 6.4 provides a checklist of normative principles and criteria for assessing climate finance in a development context. The provision of climate finance remains one of the most critical dimensions of international climate policy, with developing countries demanding that industrialized nations provide finance that is new and additional to existing aid flows, and that governance of climate finance is not dominated by richer countries.

Table 6.4 *Normative principles and criteria for climate change funding*

Delivery phase	Principle	Criteria
Fund mobilization	Transparency and Accountability	Financial contributions by individual countries and international organizations and agencies as well as their composition and sources are disclosed publicly and timely
	The Polluter Pays	Financial contributions are relative to the quantity of (historic) emissions produced
	Respective Capability	Financial contributions are correlated with (existing) national wealth and (future) development needs
	Additionality	Funds provided are more than existing national ODA commitments and are not counted towards fulfilment of existing national ODA commitments
	Adequacy and Precaution	Amount of funding is sufficient to deal with the task of maintaining global temperature rise below 2°C
	Predictability	Funding is known and secure over a multi-year, medium-term funding cycle
Fund Administration and Governance	Transparency and Accountability	Accurate and timely information on a mechanism's funding structure, its financial data, the structure of its Board and contact information for its Board members, a description of its decision-making process and the actual funding decisions made as well as the existence of a redress mechanism or process
	Equitable Representation	Board representation of stakeholders on the Board of a fund or funding mechanism
Fund Disbursement and Delivery	Transparency and Accountability	Disclosure of funding decisions according to publicly disclosed funding criteria and guidelines; duty to monitor and evaluate implementation of funding; existence of a redress mechanism or process
	Subsidiarity and National / Local Ownership	Funding decisions to be made at the lowest possible and appropriate political and institutional level

Precaution and Timeliness	Absence of scientific certainty should not delay swift and immediate disbursement of funding when required
Appropriateness	The funding modality should not impose an additional burden or injustice on the recipient country
Do No Harm	Climate finance investment decisions should not imperil long-term sustainable development objectives of a country or violate basic human rights
Direct Access and Vulnerability Focus	Financing, technology and capacity building to be made available to the most vulnerable countries internationally and population groups within countries as directly as possible (eliminating intermediary agencies where not needed)
Gender Equality	Funding decisions and disbursement take into account the gender-differentiated capacities and needs of men and women through a dual gender-mainstreaming and women's empowerment focus

Source: Schalatek and Bird, 2011.

As critical as the source of these flows is the way that they are allocated and spent such that those who have the greatest need and the greatest claim on these funds are able to benefit from them through improved well-being and reduced poverty. Neither international nor domestic public finances are likely to provide the required levels of funding suggested by estimates for adaptation and mitigation, meaning a crucial role for private finance and investment. Securing private sector investment requires a policy environment from international to national levels supportive of a low carbon and climate resilient future. To get there, we may need to re-think fundamentally our approach to future development, as explored in the final two chapters.

Summary

- Climate finance is important both to ensuring that development and poverty reduction progress is not reversed and on the basis of social justice, with richer countries and people compensating poorer nations and people who suffer the most severe consequences.

- Estimates suggest that large-scale financial resources will be required to meet the mitigation and adaptation challenges in developing countries. These resources will need to be met by a range of public and private sources.

- Existing sources of climate finance are highly diverse, but both pledges and spending remain significantly below the estimates of that needed.

- The extent to which these transfers of climate finance are additional to existing flows of international aid and private sector finance is the subject of intense debate.

- Existing and proposed climate finance can be assessed in terms of:

  o Governance – how climate finance is managed and controlled.

  o Sources – how climate finance is raised.

  o Disbursement – how climate finance is allocated and spent.

Discussion questions

- How can public finance best be deployed to leverage actions in the private sector in developing countries?

- What is the correct balance between mitigation and adaptation finance, and who decides on this balance?

- Are there trade-offs between additionality and mainstreaming of climate finance or are the two compatible?

- What measures can ensure that climate finance improves the lives of poor people as well as reducing global impacts?

- In the absence of significant and reliable sources of revenue from international climate finance, what options exist for national or local revenue-raising measures to finance climate change responses?

Further reading

AGF (2010) *Report of the Secretary-General's High-level Advisory Group on Climate Change Financing*, 5 November 2010, United Nations, New York.

Ayers, J.M. and Huq, S. (2009) 'Supporting Adaptation to Climate Change: What Role for Official Development Assistance?', *Journal of International Development*, 20.6: 675–692.

Buchner, B., Falconer, A., Hervé-Mignucci, M., Trabacchi, C. and Brinkman, M. (2011) *The Landscape of Climate Finance*, Venice: Climate Policy Initiative. (http://climatepolicyinitiative.org/publication/the-landscape-of-climate-finance/)

Haites, E. (ed.) (2013) *International Climate Finance*, Abingdon: Routledge.

Olbrisch, S. *et al.* (2011) 'Estimates of incremental investment for and cost of mitigation measures in developing countries', *Climate Policy* 11: 970–986.

Stadelmann, M., Roberts, J. T. and Huq, S. (2010) *Baseline for Trust: Defining 'New and Additional' Climate Funding*, IIED Briefing Paper, London, UK: International Institute for Environment and Development.

UNFCCC (2007) *Investment and financial flows to address climate change*, United Nations Framework Convention on Climate Change Secretariat, Bonn. Also featured as a chapter in Schipper, E.L. and Burton, I. (eds) (2009) *Earthscan Reader on Adaptation to Climate Change*, London: Earthscan.

Briefing papers in the *Climate Finance Fundamentals* series, Heinrich Böll Foundation North America and ODI, London. Available at: www.odi.org.uk/publications/5157-climate-finance-fundamentals

Websites

www.climatefundsupdate.org
Climate Funds Update: Independent website managed by ODI and HBF, monitoring dedicated multi-lateral climate funds and bilateral climate initiatives from donor pledges through to the disbursement of project financing.

www.climatefinanceoptions.org/cfo/node/189
Climate Finance Tracker: A World Bank website reporting funding by sector, sources and finance mechanism. Information on funding objectives, financing mechanism, application procedures, project types, decision-making structure, and project examples.

www.wri.org/publication/summary-of-developed-country-fast-start-climate-finance-pledges
Fast Start Finance Tracking: WRI website that tracks and reports pledges from donor countries for Fast Start finance under the UNFCCC, including what is 'new and additional' and institutional channels used.

www.theredddesk.org/countries
REDD Countries Database of a range of ongoing REDD+ activities organized by country and summarizing key information across a broad range of areas including policies, plans, laws, statistics, activities and financing.

www.odi.org.uk/programmes/climate-change/climate-finance
Climate Finance Programme at ODI. Website includes briefing notes and research papers on the role of the private sector in climate finance; climate finance architecture; and national delivery of climate finance.

http://climatepolicyinitiative.org/publication/the-landscape-of-climate-finance/
Landscape of climate finance map. Climate Policy Institute website that maps the current status of the climate finance landscape. The website also offers information on methods, sources of data, and data-gathering.

SECTION THREE
Development futures in a transformed world

Development and climate action in a changing world

The 21st century's sustainability challenge is characterized by unprecedented speed, complexity and global reach.

- ○ **Entering the Anthropocene**
- ○ **Stressed ecosystems and resource scarcity**
- ○ **A world plagued by inequalities**
- ○ **Shifting power relations in a multi-polar world**
- ○ **Science and technological innovation**

Introduction

We have seen in the preceding chapters that **climate change** poses profound new challenges for development, and will require important changes to the way development is thought about, planned and carried out. Climate change is not, however, occurring against the backdrop of a stable world: it is playing out in a world that is being radically reshaped by global economic, demographic and ecological forces. These changes further complicate development and **climate** action. In many cases, global warming exacerbates the challenges posed by these broader tectonic shifts.

This chapter considers the broader context of several other challenges besetting development in the twenty-first century. How policy makers, businesses, communities and individuals respond to these broader and often more immediate challenges will shape and impact their response to climate change.

The impact of human activity on nature has grown so great as to rival nature's own awesome forces: experts suggest that we have now entered or are on the cusp of a new geological epoch, the 'Anthropocene', marked by the emergence of humans as a dominant, planet-altering force. Never before has there been such a coupling of planetary and human systems on a global scale. We start by considering the implications of this new era.

Even without climate change, the future of development would shape up to be quite different from its past. Other ecological, socio-economic and demographic trends are altering the global development landscape in profound and uncertain ways. Many of the distinctions, definitions and assumptions upon which development orthodoxy rests are being thrown into question. For example, international development cooperation remains in large part premised on the outdated assumptions that poor people live in the poorest countries, and are predominantly found in rural areas. In reality, most poor people currently live in countries classified as middle income, and along with rapid urbanization poverty is becoming an increasingly urban phenomenon. Similarly, traditional dichotomies of state-led versus market-driven approaches to development may be constraining in a world where power is being redistributed and new forms of multi-stakeholder action are emerging.

In sum, we are navigating a changing sea using outdated maps. This chapter identifies the main global forces that will shape the future development landscape – and which may be further complicated by the impacts of climate change.

1. Entering the Anthropocene

The term 'Anthropocene' was first coined in the 1980s, and was popularized in recent years by Nobel Prize-winning atmospheric chemist Paul Crutzen and others, who noted that humans' collective influence on the biosphere – that thin crust of the Earth 10 miles above and below sea level that encapsulates the totality of life as we know it – had become so significant as to constitute a new geological epoch (Crutzen, 2002).

Although the human species has survived climatic conditions quite different from today's, the development of modern human societies occurred entirely within the geological epoch known as the Holocene, which started approximately 12,000 years ago (see Figure 7.1). The Holocene is characterized by a stable climate, which provided an environment conducive to the flourishing of our species and the emergence of civilized society. All written history, the advent of agriculture, and the development of cities occurred within the Holocene. In other words, the Holocene is the only planetary state known to support human civilization, as we know it.

Skeptics point out that prophesies of doom and ecological collapse have come and gone, yet modern human society continues to flourish (Lomborg, 2001). And indeed, humans have altered their physical landscapes throughout the entire Holocene period. So what's different now to warrant the declaration of a new epoch?

The sustainability challenge that the world now faces is unlike any other experienced throughout human history. Three factors combine to

Figure 7.1 *The Holocene: Climate for civilization.*

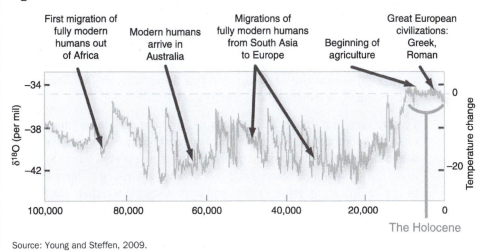

Source: Young and Steffen, 2009.

set this current era apart from its historical antecedents: (a) the global scale of human-made ecological crisis, (b) the speed and accelerating pace of environmental change, and (c) the growing interdependence of socio-economic and natural systems. Climate change is a contributing or exacerbating factor to each of these trends.

1.1 Global scale of change

The global scope and sheer scale of today's ecological crisis is unprecedented: human pressures on the environment have never been this great. Thanks to the industrial revolution, the green revolution, and advances in modern medicine and public health, the world's population skyrocketed from about 1 billion in 1800 to more than 7 billion today. It is on track to reach 9–10 billion by mid-century. Our consumption of natural resources has grown much faster than our population numbers, fueled by growing affluence and the ability to harness energy from fossil fuels.

Few locations in the world remain untouched by human influence today. The domestication of plants and animals – i.e. the modification of wild species to become more useful to us – is arguably the single most striking feature of humanity's dominion of the planet. A recent attempt by scientists to map humans' global impacts on nature indicates that already in 1995 only 17 per cent of the world's land area had escaped direct influence by humans (Kareiva *et al.*, 2007). We now divert for our own use between one-third and one-half of all the energy captured by photosynthesis on the planet, have converted more land to cropland in the 30 years after 1950 than in the 150 years between 1700 and 1850, and hold three to six times as much water in storage for human use as the amount that flows in rivers (Millennium Ecosystem Assessment, 2005; Kareiva *et al.*, 2007).

For the very first time in Earth's history, humans are altering the chemical composition of our atmosphere and oceans so much that we're disrupting fundamental earth processes at a global level (Steffen *et al.*, 2011). Climate change and its impacts (e.g. rising temperatures, ocean acidification, more violent and unpredictable weather, etc.) are just one example of this fact. Even the most remote and wildest regions of the Earth are now indirectly affected by humans' influence. As the Chief Scientist of The Nature Conservancy, Peter Kareiva, stated: 'there is no such thing as nature untainted by people . . . in the modern world, wilderness is more commonly a management and regulatory designation than truly a system without a human imprint' (Kareiva, 2010: 1). The task for conservationists, according to him, is therefore no longer to 'preserve the wild, but to domesticate nature more wisely' (*ibid*).

The pattern of industrial development pursued by rich and industrializing countries can be described as 'growing first and cleaning up later'. This practice worked as long as there was someone else's proverbial backyard to relocate to once local resources were used up or pollution became severe. Indeed, the development of the first generation of industrial powers was driven by the formation of extractive economic institutions which transferred raw materials from their colonies – located mostly in the natural resource- and mineral-rich developing world (Acemoglu and Robinson, 2012).

The legacy of this practice is discernible even today in an international division of labor within which rich countries continue to reap the benefits of industrial production, while 'outsourcing' resource depletion and environmental costs to developing countries (see Roberts and Parks, 2007). This strategy is now nearing exhaustion: we're running out of 'backyards' and of resources to extract – both in the developing and developed world. The fact that the impacts of industrial activity are so large that they're disrupting crucial global ecological processes – such as climate **stabilization**, ozone protection, the nitrogen cycle and oceanic circulation – further limits our ability to 'keep using'.

To illustrate the planetary dilemmas that define the Anthropocene, a group of scientists from around the world, led by Johan Rockström of the Stockholm Resilience Centre in Sweden, sought to identify and quantify a core set of biophysical thresholds, or 'planetary boundaries'. Within these thresholds, humans can continue to thrive; if surpassed, disastrous consequences could result for humanity. These boundaries define a 'safe operating space' for humanity, i.e. the environmental pre-conditions for **human development**. After examining numerous interdisciplinary studies of physical and biological systems, the team determined that nine environmental processes could disrupt the planet's ability to support human life. Three of these, including climate change, have already been breached (see Figure 7.2).

Figure 7.2   Planetary boundaries.

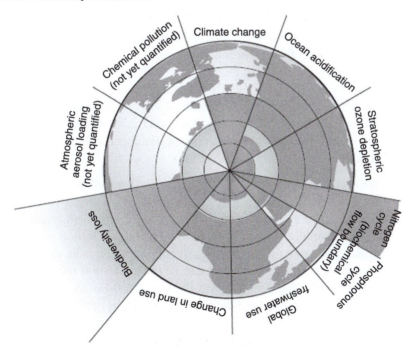

Credits: Azote Images/Stockholm Resilience Centre.

Alarming as climate change is, species loss and nitrogen pollution exceed safe limits by greater degrees. Other critical environmental processes are also heading toward dangerous **tipping points**. According to Crutzen and the planetary boundaries scientists, the global economy has grown so big that its behavior is becoming a significant factor determining the functioning of Earth systems. In the words of Steffen *et al.*, 'the human enterprise is now a fully coupled, interacting component of the Earth System itself' (Steffen *et al.*, 2011).

1.2 Speed and accelerating pace of change

The speed of anthropogenic environmental change is unprecedented, and it is accelerating: we live in an age of exponential change, yet our mindsets and management systems are still predominantly geared towards a world of linear change. A useful way to gauge the growing impact of human activity on the environment is the 'I = PAT identity', put forward in the 1970s by pioneering ecologist Paul Ehrlich. According to this simple equation, the three main factors which determine the ecological impacts (*I*) of human activity are: population size (*P*); affluence (*A*) (i.e. the level of consumption by that population); and the level of technology (*T*) used to extract resources and convert them into useful goods and wastes or pollution. The scale of environmental pressures and the speed at which they have piled on are starkly illustrated in Figure 7.3, which represents the evolution of the *PAT* factors since 1900. The figure illustrates how humanity's environmental impact increased most rapidly in the period after 1950, during which time the global population tripled and the economy grew many times faster. This period has been dubbed the 'Great Acceleration' (Hibbard *et al.*, 2006).

All signs show that the speed of change will accelerate even further in the years ahead. Whereas rich, industrialized countries were the engine of the global economy throughout the twentieth century, the big emerging economies – home to upwards of 3 billion people – have taken over since the turn of the millennium as the main drivers of global economic growth. At the same time, Africa is taking off both economically and demographically. One can therefore expect the sheer momentum behind the '*A*' factor to be far greater in the twenty-first century than it was during the Great Acceleration. Although the rate of global population growth has come down from its peak in the 1980s, global population is still set to reach 9–10 billion this century (UN Population Division, 2010). Thus, all trends point to a greater

Figure 7.3 *Evolution of the 'PAT' factors, 1900–2011 (adapted from Kolbert, 2011).*

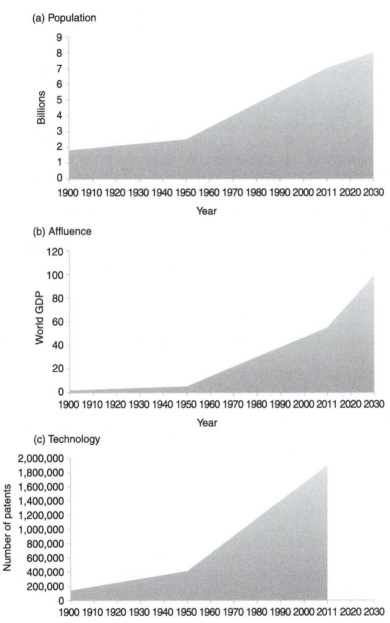

(a) Population

(b) Affluence

(c) Technology

Source: Linear averages between data points taken from Kolbert, 2011 for 1900, 1950, 2011 and from US Census Bureau (2012) and United States Department of Agriculture, Economic Research Service respectively for 2030 population and world GDP figures.

acceleration in environmental impact driven primarily by growing affluence in the emerging economies, overlaid onto existing and ongoing impacts from industrialized country affluence.

1.3 Growing complexity and resource interdependencies

With globalization, the increasing scale and velocity of change, and the ever-closer entanglement of human and earth systems, comes greater complexity. Economic, social and ecological systems are more connected than ever before – through trade, global finance, migration and the diffusion of new technologies or social innovations. This entanglement of systems creates new opportunities for shared prosperity, but it also brings heightened vulnerabilities and a greater inter-connectedness of risks (World Economic Forum, 2012). The complex interaction and interdependency of systems and events is the third aspect of the Anthropocene.

The increasing interconnectedness between our energy, water, and food systems is well documented. A landmark 2011 McKinsey report entitled *Resource Revolution* points out that the correlation between energy, water, and food prices is now higher than at any point in the past century. Water supply and food systems have become highly dependent on fossil energy inputs, which account for roughly 70 per cent of the cost of groundwater (fuel is needed to operate water pumps to extract and distribute water), and 15–30 per cent of the cost of crop production (energy-rich fertilizers are an essential input to agricultural production) (Dobbs, 2011). And conversely energy supply and distribution systems depend crucially on water resources – both directly in the case of hydropower and indirectly for cooling in the case of thermal power plants or for growing bio-energy crops.

The extent to which the global food system has become reliant on global energy and financial systems became painfully apparent with the combined food, fuel and financial crises of 2007–08 (see Box 7.1). Since the turn of the century, the average annual volatility of international resource prices has been more than three times higher than over the course of the twentieth century (McKinsey Global Institute, 2011). Resource scarcity and rising prices for key commodities disproportionately affect the poor, who rely disproportionately on the environment and natural resources for their livelihoods and well-being.

Box 7.1

Anatomy of a global systems failure: The 2007–2008 Food Crisis

The 2007–2008 Food Crisis provides a good case study in the interconnectedness of global systems and the risks it creates.

Soaring international food prices in 2007/2008 led to sharp increases in the incidence and depth of food insecurity around the world, sparking food riots in 36 countries. The poorest were affected most, as they already spend a high proportion of their income on food.

The Food Crisis occurred synchronously with rapid increases in oil prices and the onset of one of the worst financial crises since the Great Depression of the 1930s. Were these three crises linked, or was the timing coincidental?

Food prices had been climbing since 2002 as a result of changing demand and supply pressures. Increasing demand from Asia for grain-intensive meat and dairy products came at a time of tightening global grain supplies, due to years of poor harvests in the world's bread-basket regions. Global food production could not keep up with rising consumption, and global grain stores fell to historic lows. This growing scarcity helped drive up food prices.

Rising energy prices were undoubtedly an important factor in food price rises, too. Soaring energy prices pushed up the costs of food production and transportation, reflecting the energy-intensity of modern food supply chains. According to Donald Mitchell, a lead economist at the World Bank, fertilizer prices increased 150 per cent in the five years preceding 2007. These rising fertilizer costs came on the heels of rising energy prices, contributing to 15–20 per cent of the increase in US food production and transport costs over the same time period.

Evidence points, however, to the massive surge in biofuels production as the key driver behind the acceleration of food-price inflation from 2006 onwards. According to the International Monetary Fund and World Bank, the boost in biofuels contributed to increases in food commodities prices of 70–75 per cent from 2006–2008. Biofuels production took significant quantities of maize and oilseeds directly out of the food chain just as demand was trending upwards: three-quarters of the increase in global maize production from 2005–2008 went into producing ethanol. Coupled with import restrictions, biofuel mandates also resulted in inefficient uses of global crop lands. Much of the expansion of maize and oilseeds for biofuels production directly displaced wheat and soybeans.

Financial speculation and food export bans further exacerbated food-price inflation. Against the backdrop of global financial turmoil, the food-price spiral triggered intense financial speculation in food commodity futures. And as social unrest spread across the world, many countries responded by restricting food exports. Both events fueled additional food-price increases.

An important lesson from the 'triple-F' crisis is that preventing future food crises will require addressing a suite of global challenges to food security, including energy and water security, trade, and climate change. Developing more water-efficient and drought-resistant crop varieties – and deploying them in areas facing water scarcity – will help build resilience to climate change. A successful conclusion to the WTO Doha round of international trade talks would help lower volatility in world food prices, thus prompting longer-term investments and sharpening supply-side responses to price changes.

Sources: Mitchell, D., 2008; Oxfam, 2008; World Bank, 2008.

Although in many respects **vulnerability** to climate and other shocks decreases with growing affluence, we have seen that economic development sometimes increases exposures and sensitivities to such events and results in **maladaptation** (IPCC, 2012; Barnett and O'Neil, 2010).

The inherent instability of complex systems is also evident in how actions or impacts in one system can cross over into others. For instance, increasing urbanization and human mobility raises susceptibility to disease epidemics, as it multiplies the ease and speed at which diseases can spread. The impacts of natural disasters now reverberate around the globe through financial, political and communication systems.

With greater interconnectivity and cross-system interdependencies, therefore, comes a greater risk of synchronous failures of our social, economic and biophysical systems, arising from simultaneous, interacting stresses acting out at multiple levels (Folke *et al.*, 2011). A corollary of this idea is that solutions to any one problem will often depend on coordinated action in related sectors. For example, improving food security will require alleviating land and water scarcities (including challenges posed by climate change), engaging with energy policy to avoid energy crops competing with food crops, and addressing trade policies. Just as the interconnectedness of resources creates much scope for negative trade-offs, so is there potential for positive synergies.

In summary, the increasing scale, velocity and complexity of change raises the stakes for humanity and calls for urgent coordinated action across different spheres, especially as we may be reaching critical risk thresholds, or '**tipping points**'. An important implication is that **resilience**, adaptability and learning must be integral to development and climate interventions (see Chapter 5).

2. Stressed ecosystems and resource scarcity

2.1 The biodiversity crisis

Closely linked to climate change, rapid and accelerating species extinction and ecosystem degradation is arguably the most pressing planetary concern. The rate of species loss is greater now than at any other time in human history. Extinctions occur 100 to 1000 times faster than the natural extinction rate (Foley, 2010). We're now experiencing the sixth episode of mass species extinction that our planet has known (the last one occurred 65 million years ago, when dinosaurs and many other species were wiped out).

Previous periods of mass extinction and ecosystem change were driven by global changes in climate and atmospheric chemistry, e.g. asteroid impacts or volcanic mega-eruptions. This time, the main driver is the growing competition for resources between us humans and all other species. The main cause of extinction is habitat loss or degradation, chiefly through the felling of tropical rainforests – the most diverse ecosystem – for the expansion of industrial agriculture.

Biodiversity matters to humans in many more profound ways than our caring about the predicament of pandas or Bengal tigers:

- Biodiversity underpins our entire food supply chain and is therefore crucial to food security: for example, kelp forests and coral reefs provide nursery areas and feeding grounds for the fish we eat, and we rely on bees and wild birds to pollinate our crops and orchards (Thrupp, 2000). According to the United States Department of Agriculture, bees pollinate 75 per cent of the nuts, fruits, and vegetables growing in the United States. Moreover, in the context of a changing climate, the wild relatives of crop plants are an increasingly important resource for improving agricultural production and for maintaining resilient and sustainable agro-ecosystems. We depend on these wild stocks to protect our agricultural stocks from future damage or loss.
- Biodiversity is vital to our health (Johns and Eyzaguirrea, 2007). Most of our medicines come from nature. More than two-thirds of people living in sub-Saharan Africa use traditional herbal and plant medicines for primary health care (WHO). Conserving

biodiversity is therefore important both for current health care and as a form of insurance against future illness and diseases.

- Biodiversity contributes to the stability, resilience and productivity of ecosystems. In nature's intricate web of life – characterized by the interdependence of species – no species is redundant, no matter how small. All play a role in the functioning of ecosystems. Nature's genetic diversity is a precious source of resilience for ecosystems and the people/communities that depend on them.
- Biodiversity contributes greatly to global economies and livelihoods. The Convention on Biological Diversity – officially adopted by 193 countries – notes that 'at least 40 per cent of the world's economy and 80 per cent of the needs of the poor are derived from biological resources'. The pharmaceutical, biotechnology and agricultural sectors – which are important pillars of economic activity in the world economy – depend directly on bio-prospecting of wild species for future production and growth. Biodiversity is therefore an essential raw material for continued discovery and wealth creation.

Alarmingly, evidence suggests that there has been no reduction in the rate of biodiversity loss over the past four decades, and the rate of human response to this loss has slowed over the past decade (Butchart *et al.*, 2010; in Steffen *et al.*, 2011).

Biodiversity loss can be seen as one effect of the broader problem of ecosystem degradation. In 2005, the UN-led Millennium Ecosystem Assessment (MA) – the first scientific, global stock-taking of the world's ecosystems and the services they supply to people – revealed that two-thirds of global ecosystem services are already degraded due to human activity.

Human activity over the past five decades in particular has transformed ecosystems on a scale that dwarfs the cumulative impact of human history. The most significant change has been the transformation of large chunks of Earth's terrestrial surface to cultivated systems. More land was converted to cropland in the 30 years after 1950 than in the 150 years between 1700 and 1850 (Millennium Ecosystem Assessment, 2005). According to the World Resources Institute, more than 80 per cent of the Earth's natural forests already have been destroyed, and up to 90 per cent of West Africa's coastal rain forests have been destroyed since 1900.

The human impact on marine ecosystems has been similarly dramatic. Approximately 35 per cent of mangroves – critical habitat for supporting marine life and protecting coastlines from erosion – have been lost in the last two decades. Today, 90 per cent of the world's large fish – such as tuna, swordfish, and marlin – have disappeared as a result of overfishing. Roughly 20 per cent of the world's coral reefs were lost and an additional 20 per cent degraded in the last several decades of the twentieth century (Millennium Ecosystem Assessment, 2005).

The most rapid ecosystem changes are now taking place in developing countries, although in the past, industrial countries experienced comparable rates of change. Unsurprisingly, changes prior to the industrial era seemed to occur at much slower rates than current transformations (*ibid*).

Climate change has a significant and exacerbating effect on biodiversity loss and ecosystem degradation. As global temperatures warm, many species will be driven pole-wards or towards higher altitudes in pursuit of cooler climes. Many of the world's biodiversity hotspots are located at the edges of land masses (e.g. in Madagascar, South Africa, southern Australia) so as temperatures warm, species in those locations will have nowhere to go. Large-scale extinctions will ensue. Thus, many plant and animal species are unlikely to survive climate change.

Healthy ecosystems also play an important role in regulating the climate. For example, terrestrial and marine ecosystems have provided an immense ecosystem service to humanity over the past 150 years by absorbing close to 50 per cent of global carbon dioxide emissions (Millennium Ecosystem Assessment, 2005). Damage these ecosystems and you eliminate important carbon sinks, fueling climate change and spurring continued ecosystem degradation.

The Millennium Ecosystem Assessment (MA) predicts that in the course of the twenty-first century, climate change will emerge as the dominant driver of ecosystem degradation. The degradation of natural capital such as fisheries, forests and watersheds places these important systems at risk of collapse or extinction. Losing them will jeopardize our ability to feed, house and clothe a growing population, to say nothing of the livelihoods that depend on these ecosystems. Rural livelihoods in particular risk being undercut by ecosystem loss, as half of all jobs worldwide depend directly on agriculture, forestry and fishing.

Thus, beyond saving animals and habitats, protecting biodiversity and managing ecosystems is about preserving the very foundations of global economies and local livelihoods. It also helps us mitigate climate change and cope with its impacts. Conversely, climate action can help limit further ecosystem degradation and help avert dangerous climate tipping points.

2.2 Resources scarcity and interdependencies

Managing resource scarcities and ensuring food and water security are likely to be persistent global challenges in the face of mounting supply-side and demand-side pressures on natural resources. Global resource consumption is set to rise as the global economy expands in tandem with population increases and the rapid growth of emerging economies.

Over the course of the past half century, the global consumption of food and fresh water has more than tripled, while fossil fuel use has risen four-fold. The global economy is set to continue on its current growth trajectory, implying a doubling in size every two decades. As the global population surges towards 9 billion and 3 billion more consumers join the ranks of the middle class by 2030, the pressure on resources will be tremendous. According to the McKinsey Global Institute (Dobbs, 2011), growing populations and rising wealth will increase the demand for all major natural resources (energy, water, fiber, food) by 30–80 per cent by 2030. How to meet that demand without compromising exhausting the Earth's resources and critical ecosystems is a colossal challenge. Concurrently, rampant ecosystem degradation and worsening climate change pose profound challenges for the supply of key commodities.

These are not distant threats: there are clear signs that the resource crunch is already upon us. In 2012, an estimated 2.7 billion people were affected by water shortages. That figure is set to increase: global demand for fresh water is on track to outstrip supply by 40 per cent in the next two decades (Comprehensive Assessment of Water Management in Agriculture, 2007). It is estimated that by 2025, up to two-thirds of the world's people are likely to live in water-stressed conditions (*ibid*). The latest edition of the World Economic Forum's Global Risks Report ranks water supply among the top five global risks, on a par with systemic financial failure and fiscal imbalances (World Economic Forum, 2012).

Plate 7.1 *Atlas of global water stress: (a) Current water stress; (b) Projected change in water stress to 2025 due to climate change.*

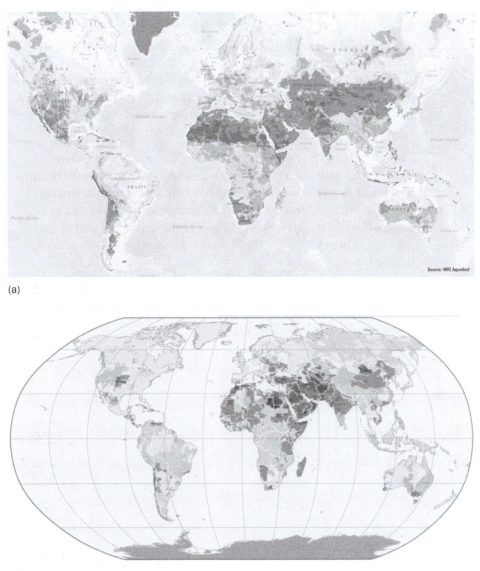

(a)

(b)

Credit: World Resources Institute (WRI).

Note: These maps are produced by WRI using the 'Aqueduct' tool which provides users with the highest-resolution, most up-to-date data on water risk across the globe. The dark sections in plate 7.1(a) indicate areas currently facing water stress (the darker the shading the greater the degree of stress). This shows that the most water-stress regions are also among the most populated; in particular China and South Asia face severe water scarcities. Plate 7.1(b) indicates that these regions are also among the worst affected by climate impacts. It shows projected changes in water stress to 2025, factoring in the impacts of climate change (using IPCC projections). The darker the shading the bigger the increase in water stress. An estimated 1.8 billion are projected to live in areas with severe water scarcity by 2025. An interactive atlas can be accessed at: http://aqueduct.wri.org/atlas.

At the same time, food productivity increases are stalling: the world consumed more food than it produced in seven of the eight years between 2000 and 2008, resulting in global food stockpiles reaching historical lows. Food demand is projected to rise by 50 per cent by 2030 (World Bank, 2010a). And rising and increasingly volatile prices for food, energy and minerals over the past decade have reversed the previous century's trend toward cheaper resources (see Figure 7.4 below). We'll not only face strained natural resources in the coming years – we'll face escalating prices for what resources we do have.

The situation with energy and atmospheric resources is equally worrisome – particularly in emerging economies. China already ranks as world's largest emitter of carbon dioxide from fuel combustion. Over the next two decades, China will become the world's largest energy consumer: it is on track to use 70 per cent more energy than the United States by 2035 (IEA, 2011). The fastest growth in energy demand, however, comes not from China, but from India, Brazil, Indonesia, and the Middle East. On current trends India – which is

Figure 7.4   *Rising international commodity prices.*

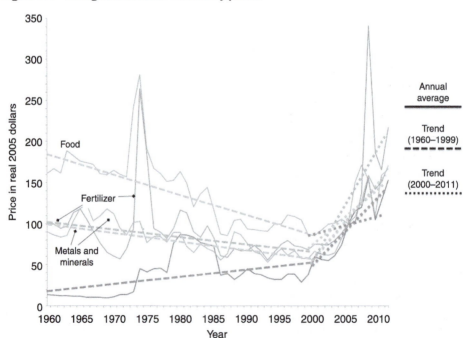

Source: World Bank data; Credit: WRI, with thanks to Aaron Holdway.

expected to overtake China as the world's most populous nation by 2021 – is expected to more than double its demand for coal by 2035 (United Nations Population Division, 2010; OECD/ IEA, 2011). These emerging countries will also drive a huge increase in car ownership. As of 2010 there were about 800 million vehicles in use; by 2050, that number will rise to 2.5 billion, with most of the increase coming from the developing world. Together, China, India and Brazil added about 20 million cars to the road in 2010 alone (*ibid*). We will need to figure out a way for these economies to meet their growing resource demands in ways that don't tip humanity past critical environmental risk thresholds, whilst at the same time reducing the environmental and resource impacts of the already industrialized nations.

So how worried should we be?

In his 1983 book *The Ultimate Resource*, libertarian economist Julian Simon famously argued that resource scarcities are always temporary conditions that scientific and technological advances eventually overcome. The classic economic argument put forward in this book said that scarcity should be understood in economic rather than physical terms. As far as humans are concerned, physical resource limits are irrelevant: before resources run out, *economic* scarcity – conditioned by market forces and signaled through price increases – will create the necessary incentives for human ingenuity (a limitless resource) to develop substitutes, expand production frontiers, increase efficiency or ration and recycle.

Indeed, history shows that impending resource scarcity can drive technological innovations that increase the efficiency of resource use, spur the development of substitutes, and improve resource extraction capability (such as shale gas and oil sands).

Despite the potential for innovation, however, several factors may thwart future efforts to address resource scarcity:

- Global scale: nearly all accessible food, water, and fossil resource zones are now in production, with the exception of critically precious ecological zones such as the tropical rainforests of the Amazon, Congo Basin and Sumatra, and extremely hard-to-reach areas such as the Arctic and ocean bottom (Klare, 2012).
- Demand rebound: efficiency improvements can, paradoxically, drive net increases in demand. People may drive more miles when their cars are more energy efficient, for example. This is known as the 'rebound effect' (Sorrell, 2007).

- Limitations on substitutes: substitutes can take a long time to mature, and even then it is not certain that alternatives will be able to substitute on a one-for-one basis.
- Leakage: the development of substitutes can increase pressure on resources that are already in high demand. Almost all the vital minerals that we depend on – including cobalt, nickel, and titanium – are already facing supply challenges (Klare, 2012).
- Non-physical and non-economic constraints: this includes the imperative to safeguard ecosystems and protect indigenous peoples' rights in frontier resource rich areas.
- Climate change: climate change is already contributing to water scarcity and food insecurity.

The interplay between climate change and resource scarcity is ambiguous. Global warming may open new resource frontiers, as is the case with the melting of the Arctic ice. However, the imperative to reduce **greenhouse gas** emissions may limit the continued exploitation of fossil resources, especially emissions-intensive 'unconventional' hydrocarbons like tar sands and shale gas. Climate change and resource degradation are causally related and mutually reinforcing: climate change is a direct driver of biodiversity loss and ecosystem degradation and, conversely, the loss of ecosystems such as rainforests and peat-lands contributes to climate change.

Continued strain in commodity markets may portend a new kind of resource geopolitics. Many countries are already responding to the resource crunch by seeking to secure reserves abroad, purchasing land or investing in the expansion of extractive industries. According to Oxfam more than 80 million hectares of agricultural land have been acquired by foreign investors over the past decade, two-thirds of which was in sub-Saharan Africa (Oxfam, 2012). To put it in perspective, that amounts to an area the size of Spain, France, Great Britain, Italy and Germany combined.

What is true for agricultural land is no less true for other natural resources. In his 2012 book *The Race for What's Left*, Prof. Michael Klare describes how unprecedented shortages in uranium, copper, vital minerals such as bauxite, cobalt, nickel, titanium and platinum, and rare earth elements, are driving a frenzied race by governments and corporations to control and exploit remaining resources (Klare, 2012). Commodities now account for about two-thirds of foreign direct investment by state-owned enterprises (National Intelligence

Council, 2012). Another alarming manifestation of intensified competition for resources is the rise of 'resource nationalism', when governments assert control over the natural resources located in their territories (e.g. through export bans). This is a trend that the international system is ill-prepared to address, as no effective international frameworks currently exist for dealing with export controls, which often exacerbate resource shortages.

The resource crunch and its manifestations in intensified competition and rising commodity prices will disproportionately affect poor people. As we have seen in Chapter 2, impoverished communities tend to rely heavily on the environment and natural resources for their subsistence and livelihoods. They are therefore acutely vulnerable to environmental resource depletion and degradation. Rising prices for food, fuel and land will have severely regressive impacts – not least since poor people spend a much higher proportion of their income on energy and food. The effects of the 2007–08 Food Crisis provide a stark reminder of this, plunging 44 million people into poverty (FAO, 2011). And the scramble for resources may lead to greater capture and control of these resources by the powerful, to the detriment of the poor. Even where legal frameworks are in place to ensure a certain level of resource access, poor people may find their rights trampled – for example, through forced displacement from land – as scarcity increases. According to noted science journalist Fred Pearce, land grabs may have a bigger impact on the world's poor than climate change for that reason.

3. A world plagued by inequalities

Globalization has produced winners and losers. While there has been a degree of income convergence between rich and poor *countries* in recent years – due chiefly to the stellar performance of the emerging economies – income inequality between *people* has increased. Inequality is growing both within and between countries at the extremes of the wealth spectrum (IMF, 2007; Berg and Ostry, 2011b). As revealed by the global consultations conducted to inform the post-2015 development goals, there is a widespread perception that inequality and unfairness are at unacceptably high levels, and that this is creating social stress.

Despite six decades of global economic growth that followed the end of the Second World War, the gap between rich and poor countries has stubbornly persisted: the same group of countries has remained at the top of world-income distribution, while only a handful of countries that started out poor have joined that high-income group (UNDP, 2011). Countries in the top 10 per cent of the global income list are now about 100 times richer than the bottom tenth. While a small percentage of the global population has accumulated vast wealth, billions have been left behind: they are experiencing poverty, joblessness, hopelessness and falling living standards. In 2011, the world's 1,400 or so billionaires, constituting 0.0000002 per cent of the world's population, earned the same amount as the combined incomes of the poorest 1.9 billion people who together make up more than a quarter of the world's population. One in eight young people globally are without jobs and out of school. In most countries, even where national incomes are growing, the gains are concentrated.

While the emergence of a 3 billion strong global middle class will help close some of the income gap, income inequalities will remain high in many countries. And other forms of inequality may persist or widen. Inter-generational inequalities are expected to emerge as an issue in countries with ageing populations, where the young will have to pay the bill for rising pension and health-care costs of the elders. Rapid urbanization and the rise of megalopolises in the developing world will likely deepen the rural–urban divide, as cities suck up talent and resources, leaving rural areas depressed and facing long-term decline. This has been a concern for example in China, where the government has taken strong measures to counter this trend by encouraging the development of the country's rural hinterland.

The most invidious form of inequality is inequality of opportunity, when a person's life chances are largely determined by the circumstances of their birth. A child born into a poor household in sub-Saharan Africa is two to three times less likely to receive a primary education than a child born into a wealthy household in the same country. That child is about 18 times more likely to die before the age of five than a peer in a high-income country. Being born in the city as opposed to rural areas doubles a child's chance of reaching five. An astounding 60 per cent of a person's income is determined merely by the location of her birth, with an additional 20 per cent dictated by how rich her parents were (Milanovic, 2010).

The trend lines are not reassuring. By US Census Bureau projections, more than 40 countries – including many African countries, Central

Asian states, and Russia – have a lower life expectancy in 2010 than they did in 1990 (US Census Bureau, 2012). Since the 1980s, more than twice as many countries have seen income disparities widen than have seen them narrow (UNDP, 2011). Even within Asia, Gini coefficients – which are standard measures of inequality – are showing a growing gap between the rich and the poor, particularly between rural and urban populations (*ibid*).

Some of the factors that contributed to the global increase in inequality include the wave of deregulations brought about by the Thatcher and Reagan governments of the 1980s, the collapse of communist systems in Eastern Europe and Russia, and the opening and liberalization of the Chinese economy. Latin American countries have bucked this general trend: long the region with the widest income and wealth disparities, income inequalities have fallen due to a mixture of more progressive public spending and targeted social policies (e.g. cash transfers).

In the mid 1950s, economist Simon Kuznets postulated an inverted-U-curve relationship between economic development and income inequality. Under this model, inequality increases initially and then declines as economies develop and the majority of the workforce shifts from low-productivity sectors into high-productivity sectors (Kuznets, 1955). The evidence of rising inequality in the developed world, however, invalidates Kuznets' hypothesis of an automatic U-curve relationship between economic growth and inequality. Income inequality is on the rise throughout OECD countries, as well as in Asia. The United States is now the most unequal country amongst the advanced industrial economies.

There are many reasons why inequality matters to poverty reduction and international development (see Chapter 2). Not least among these is that rising inequality reduces the effectiveness of pursuing economic growth as a method of reducing poverty. There is growing evidence that excessive inequality damages long-term growth prospects, perhaps because it shrinks the middle class. Inequality is also linked to greater economic instability and risk of conflict (Berg and Ostry, 2011b). A report by the Brookings Institution shows that poverty reduction can be much faster if the rise in inequality is held in check (Chandy and Gertz, 2011).

Inequality also matters for international action on climate change. As Roberts and Parks explain, 'the actual and perceived inequality between nations also creates starkly disparate worldviews and a poisonous mistrust that makes it impossible to reach the ambitious cooperative agreement needed to address climate change effectively' (Roberts and Parks, 2007: 216). Of more immediate concern, finding an equitable means of sharing out the available global **carbon budget** remains a prerequisite for any genuinely comprehensive, international agreement for managing climate change.

There are signs that growing inequalities are rising up the global agenda. In 2011, business leaders at the World Economic Forum identified wealth inequality as the most serious challenge facing the world in the years ahead. They issued a warning that the gap between the rich and the poor within both developed and developing nations needs to shrink in order to build a more sustainable economy. The Arab Spring and the Occupy Wall Street movement in the United States can be seen as movements for justice in response to growing inequalities.

Indeed, addressing inequality as a means of advancing human development is one of the 'big ideas' to emerge from the development community in recent years. This line of thinking is already reshaping development discourse and practice. Refocusing on **equity** has been a central plank of UNICEF's strategy and is becoming a guiding principle in many other development agencies. Inequality was the central theme of the 2011 Human Development Report. And there is a growing chorus demanding a more explicit emphasis on inequality in the next set of global development goals, which would take the place of the current Millennium Development Goals after 2015.

The context of growing resource scarcity will bring practical urgency to the equity and inequality agenda. As Alex and Jules Evans explain 'the emergence of natural resource limits (even if only temporary) is a fundamental game-changer for political agendas on equity and fairness, and [. . .] these considerations may increasingly come to be seen as a new front line for international development over the next decade and beyond' (Evans and Evans, 2011: 4). In the context of biting resource scarcities, 'equity and justice arguably cease to be a merely normative agenda in a world of finite resources and high interdependence – instead becoming a basic design principle for institutional effectiveness and sustainable resource management' (*ibid*).

4. Shifting power relations in a multi-polar world

Global power dynamics are shifting, bringing profound changes to international development and giving rise to a more complicated geopolitical landscape.

In the past five years, the economic and financial strength of the emerging economies – China, India and Brazil in particular – has grown, even as economies in industrialized countries have contracted. This has begun to be reflected in a realignment of formal power structures. For example, we've recently seen a modest re-distribution of voting shares in international financial institutions and the transcendence of the G-20 over the G-8 as the major forum for political and economic leadership. In the international policy-making processes, including the climate change negotiations, the role of and expectations for emerging economies designing and participating in the **regime** have grown dramatically over the past few years. The rising influence of emerging economies can be expected to continue. The US National Intelligence Council projects that by 2030 Asia will have overtaken the US and Europe combined in terms of global power, as measured by GDP, population size, military spending and technological investment (National Intelligence Council, 2012).

By 2030, the global middle class will explode to 5 billion (from 2 billion today), 66 per cent of whom will live in Asia. Two decades ago, sustainability centered on the middle class in North America and Western Europe. In the coming decades, habits of the middle class in emerging economies will shape the future of sustainability in a way that they never have before.

Urbanization is another major trend that is reconfiguring power relations, with profound implications for the future of development and climate action. For the first time in human history, the majority of the world's population lives predominantly in cities. This trend is set to continue: 70 per cent of people are expected to be living in cities by mid-century, representing a near doubling of the urban population from 2010 levels (United Nations Population Division, 2010). Much of this urban growth will occur in river basins, deltaic systems and coastal areas that are already vulnerable to climate-related **hazards**.

Cities now account for three-quarters of global energy consumption and a similar share of global carbon dioxide emissions. Rising and related problems of congestion, pollution, and inadequate city services affect the productivity and health of all, but fall particularly hard on the urban poor. Although the majority of poor people – particularly the chronic poor – are still to be found in rural areas, a massive rural–urban transition is underway. The future of poverty will thus become increasingly urban, as growing numbers of poor people find their way to cities. In urban areas poverty is characterized by unsafe housing and sanitation, high transport costs, and lack of access to energy and to other basic services (UNEP, 2011).

As an increasing percentage of the world's population moves to an urban environment, the authority and capacity of city institutions to shape the relationship between environment and development has grown. Groups like the C-40 network of 58 of the world's megacities are taking the lead in developing more sustainable policies on transportation, building codes, and water and energy service delivery. Initiatives led by these groups could begin to influence both national and global policies. They could be the harbinger of an age in which cities play a more proactive role in resources management, environmental standard setting and social welfare provisioning.

The evolving geopolitical landscape is prompting profound changes in development architecture and development cooperation approaches. The emergence of a 'Global South' – especially in the form of an assertive, purposeful voice in international affairs from big emerging powers like Brazil, China, India, Indonesia, Mexico and Turkey – has given new impetus to the call for more inclusive mechanisms of global governance. This is evident for example in heated debates about the reform of multi-lateral institutions and in deliberations around the shaping of a new international development agenda after the expiry of the MDGs in 2015.

The increasing prominence of South–South exchanges and financial flows as well as the proliferation of actors in the development arena – such as the increasing influence of private philanthropic organizations such as the Gates Foundation – has ushered an end to the oligopolistic model of development cooperation that had prevailed historically. At the same time, new instruments and mechanisms for development cooperation have emerged, channeling new forms of financing, technology transfer, capacity development and so forth. All this has made it urgent to consider new forms of cross-boundary solidarity and collective action.

At the same time, innovations in information and communications technology (ICT) will assist a broad diffusion of power to citizens and non-state actors (see Figure 7.5). Empowered by access to an

Figure 7.5 *Innovations in ICT create opportunities for empowerment.*

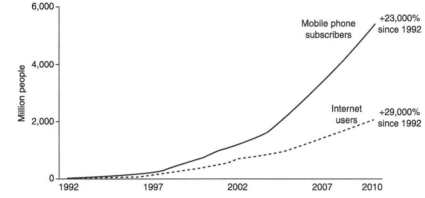

Source: World Bank ITU.

increasingly interconnected world and unprecedented capabilities to organize and collaborate, a growing number and diversity of people will be able to express their views in national and global political processes. The ICT revolution makes possible a shift in power towards 'multifaceted and amorphous networks composed of state and nonstate actors that will form to influence global policies on various issues' (National Intelligence Council, 2012). The past five years has seen this play out in cyberspace, as exemplified with the global online conversations on the post-2015 development goals (www.worldwewant2015.org), and the meteoric rise of global web-based civic movements such as Avaaz, and Iran's Twitter Revolution. The ICT revolution also puts tremendous power in the hands of corporate titans of the digital economy such as Google and Facebook which, with superior knowledge of individuals' motivations based upon unparalleled access to vast streams of real-time data, can rival state actors in their ability to affect citizens' behavior, on as large a scale.

5. Science and technological innovation

The advancement of science and innovation is a double-edged sword. It will need to be wielded, but with caution, in cutting a path to sustainable prosperity.

On the one side, there is no question that disruptive technological innovation is our best hope to defy the Malthusian curse, overcome resource limitations and reconcile climate stabilization and development goals. The arithmetic of the Kaya Identity (see Chapter 4) makes this point clearly with regard to needed carbon emissions reduction.

The world is in the midst of a science, technology and information revolution. Progress in biotechnology, nanotechnology, materials and information technologies has accelerated, aided by the exponential increase in computing power. These technological innovations have brought about an extraordinary enhancement of human well-being, as well as longer, better lives for many of the world's people. Future trends will be marked not just by advancements in individual technologies, but by a force-multiplying convergence of multiple technologies.

The ICT revolution has spurred the rise of 'big data' marked by massively improved data-collection capabilities and options – via satellite imagery, geo-coding, crowd-sourcing – and an unprecedented ability to convert into data and quantify and analyze many aspects of the world that hadn't been properly measured before, such as trust in institutions, social capital, or feelings of satisfaction with public-service delivery.

Big data matters big time because it increases transparency and access to information, which is a keystone for greater accountability and it assists a more effective design and implementation of policies and interventions: better data and statistics will help governments and businesses to track progress and incorporate lessons learnt from implementation, which will make them effective. Developing countries have a huge opportunity to leapfrog to fine-grained data collection systems that can be useful for policy making. For example, the first assessments of living conditions in South Sudan were made possible using SMS technologies, as the newly formed country had no capacity for conducting formal statistical surveys.

On the other side, the scientific and technological benefits of globalization are unequally distributed, and have not yet reached far enough. About 2.5 billion people live on less than US$2 a day, suffering from deprivation and disease. Research and development in the health and agricultural sectors remain heavily skewed towards the wealthy nations, and fail to consider the most pressing issues or needs in poorer countries, or offer solutions that are not affordable or that depend on developed infrastructures that do not exist in developing countries.

Plate 7.2 *Global Forest Watch 2.0: Technology and big data to the rescue of forests.*

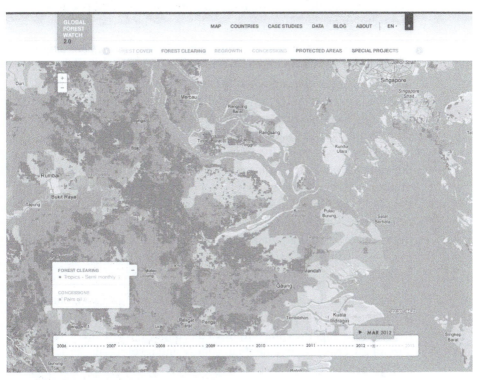

Credit: World Resources Institute (WRI).
Note: This screenshot shows real-time forest-clearing activity overlaid with maps of palm oil concessions in Indonesia. It is just one example of the web-based Global Forest Watch (GFW) 2.0 tool in action. GFW 2.0 is a powerful, near real-time forest monitoring system that unites satellite technology, data sharing, and human networks around the world to fight deforestation. It combines various near real-time, tree-cover loss alert systems, complementary satellite imagery and monitoring systems, a treasure trove of maps, mobile technology, and a networked world to create never-before-possible transparency on what is happening in forests everywhere. This transparency will enable governments, communities, civil society, companies, and the media to hold those responsible for forests accountable at a pace that matches the modern world and the threats facing forests.

The march of technology can also entail significant social costs and sometimes worsen the plight of poor people: automation and the shift from small-scale artisanal to large-scale commercial (i.e. from labor-intensive to capital-intensive) production techniques can cause big job losses in important sectors, such as agriculture. These costs usually disappear over time: historical experience suggests technology is pro-poor in the long run, and helps level the playing field by widening peoples' access to powerful tools.

The challenge for the future is to accelerate poverty reduction and enhance human well-being, while repairing and enhancing the ecosystems that create wealth. Addressing this challenge will require better, smarter globalization that relies heavily on advances in science and technology.

A host of information and communications technologies are emerging that can be game-changing in that regard. Low-cost environmental data collection using high-resolution satellite mapping and remote sensing is one example. For example, increasingly low-cost and accessible technologies are beginning to measure trends in deforestation, soil erosion and climate change globally and with remarkable granularity. The World Resources Institute (WRI) has developed a powerful, near real-time forest monitoring system that unites satellite technology, data sharing, and human networks around the world to fight deforestation (see Plate 7.2). Rural Indian farmers can receive online updates about market prices and weather on their mobile phones, making them more competitive. Emerging economies like India, China and Brazil have launched their own satellites and are sharing data with other developing countries.

Science and technology hold great promise for navigating a changing world, but we must proceed with caution. Technological prowess also expands our appetites and amplifies our impact. While there has been a trend towards improvements in energy and resource-use efficiency as incomes increase, the efficiency effect is almost always trumped by increased consumption. In the final analysis, it doesn't matter that we're more efficient if we produce and consume more. And while big data can boost transparency and democratic governance, it can also be wielded by authoritarian regimes to more closely monitor and control its people, or by corporations to influence consumer behavior. Development practitioners will need to walk a delicate line to embrace innovative scientific and technological solutions while deterring environmental degradation.

Conclusion

Climate change interacts in complex ways with other mega-trends, such as ecosystem degradation, resource scarcity, demographic shifts, shifting power relations and rapid scientific and technological innovation. Responding effectively to climate change requires an understanding of how these broader trends shape realities on the ground. Given the nature of climate change as a risk-multiplier, its effects on development cannot be fully grasped and addressed without reference to the broader context of global risk and change. This chapter considered this broader, global context and its implications for development. In the next chapter, we consider how new kinds of thinking are taking shape to help guide development efforts in the age of the Anthropocene.

Summary

- Our impact on nature has grown so large that we may have entered a new geological epoch – the 'Anthropocene' – marked by humans' dominating influence. This new age brings with it new development challenges, which will also have implications for how we respond to climate change.

- Growing resource scarcity and ecosystem degradation worldwide puts development at risk and provides greater impetus and urgency to address climate change.

- Issues of fairness and **social justice** are rising up the development agenda as the gap between rich and poor widens.

- Power shifts pose challenges for multi-lateralism and open up new spaces for multi-stakeholder action.

- Science and innovation raise new possibilities and challenges for sustainable development.

Discussion questions

- What evidence is there to indicate that we are now living in the 'Anthropocene'? What are the implications for development?

- Does the existence of planetary boundaries imply limits to growth and a dead end for development?

- What are the major forces reshaping the development landscape in the twenty-first century? How do these forces shape our responses to the climate challenge?

- Humans have proved exceedingly ingenious in innovating their way out of looming resource crunches in the past. What may be different this time?

● In what ways is power diffusing in the international system? What are the implications for international collective action?

Further reading

National Intelligence Council (2012) *Global Trends 2030: Alternative Worlds*, Washington DC: US Government Printing Office, November.

Rockstrom, J. (2009) 'A safe operating space for humanity', *Nature*, 461 (September).

Steffen, W., Persson, Å., Deutsch, L., Zalasiewicz, J., Williams, M., Richardson, K., Crumley, C., *et al.* (2011) 'The Anthropocene: From Global Change to Planetary Stewardship', *Ambio*, 40.7: 739–761.

Sumner, A. (2012) 'Where do the Poor Live?', *World Development*, 40.5: 865–877.

World Economic Forum (2012) 'Global Risks 2012 – Seventh Edition'.

Websites

www.igbp.net
The website of the **International Geosphere-Biosphere Programme (IGBP)** provides links to latest research and a wealth of resources on global-scale and regional-scale interactions between Earth's biological, chemical and physical processes and their interactions with human systems. Founded in 1987, IGBP's vision is to provide essential scientific leadership and knowledge of the Earth system to help guide society onto a sustainable pathway during rapid global change.

www.millenniumassessment.org
The **Millennium Ecosystem Assessment** assessed the consequences of ecosystem change for human well-being. From 2001 to 2005, the MA involved the work of more than 1,360 experts worldwide. Their findings provide a state-of-the-art, scientific appraisal of the condition and trends in the world's ecosystems and the services they provide, as well as the scientific basis for action to conserve and use them sustainably.

http://www.resourcesfutures.org/#!/introduction
Chatham House's 'Managing Resources Futures' explores the shifting global political economy of key resources (land, water, energy, minerals and food) through analyzing their inter-linkages in production, use and trade. Also known as the Royal Institute of International Affairs, Chatham House is a reputed British think tank that has done some work on resource issues. The website of the Chatham House report *Resources Futures* (Lee *et al.*, 2012) features an interactive digital tool that shows the emerging economies that have become major centers of resource consumption, joining existing economic powers. The tool also graphically illustrates global interdependencies; the concentration of production in a handful of countries; new producers set to join the scene; and the new wave of consumers.

In addition, there are examples of likely political, economic and environmental disruptions.

http://insights.wri.org

WRI Insights is part of WRI's mission to provide unbiased, expert analysis on the most important environmental issues facing the world today. Insights aims to be a platform for timely, honest, practical information, and the exchange of ideas and solutions. Over time, this community of contributors will grow to include partners and stakeholders on a range of subjects.

www.globaldashboard.org/

Global Dashboard is a blog focused on global risks and international affairs, bringing together authors who work on foreign policy in think tanks, government, academia and the media. It was set up in 2007 and is edited from the UK by Alex Evans and David Steven.

Alternative development futures: Pathways to climate-smart development

Economic growth measured by Gross Domestic Product offers a flawed guide for development progress.

Introduction

Climate change and the major shifts described in the previous chapter have brought into sharp focus the risks and pitfalls associated with current development paths, and thrown into question the merits and long-term viability of standard, GDP-centric approaches to development. At the same time, the global and interconnected food, fuel and financial crises of recent years have bred widespread dissatisfaction with conventional economics and given credence to the view that achieving sustainable prosperity for all requires breaking with prevailing modes of production and consumption which are exhausting the planet and straining societies, and threaten to destabilize the Earth's life-support systems.

Notions of a 'green economy', 'green growth', 'steady-state economy', 'limits to growth' and 'planetary boundaries' are resurgent in international development discourse. These share the premise that past successes can no longer provide a reliable guide to future progress.

The need to change course has never been clearer, yet all the environmental trend lines continue to point in the wrong direction. Indeed, we are accelerating in the wrong direction. Like the **climate** system, human society is subject to deep-set inertia. This inertia stems in part from human psychology (e.g. deferred responsibility and time discounting), in part from the rigidity and fragmentation of our institutions, which subdivides knowledge and action into different silos and specialities. To a significant degree, human inertia comes down to how our behaviors and decisions are governed by defunct thinking. Keynes captured this insight with his famous remark that society is controlled by the ideas of long-dead economists and philosophers: 'Practical men, who believe themselves to be quite exempt from any intellectual influence, are usually the slaves of some defunct economist' (Keynes, 1964: 383). The first task therefore is to recognize and challenge inherited ways of thinking, the assumptions and biases that inform our worldviews, in order that these may give place to more age-appropriate thinking and action.

This chapter starts by recapping the case for a re-think of development in the twenty-first century. It then briefly examines major alternative development paradigms from the perspective of consistency with climate action and planetary boundaries. While there is a range of views about the depth and scope of change needed, there is considerable agreement about the basic conditions and first steps that need to be taken on the path towards a more just and prosperous future, taking into account climate change and biting resource constraints. A final section – the bulk of this chapter – then concerns itself with the essential ingredients of this transition.

1. Development at a crossroads

Just as a glass filled halfway can be described as half empty or half full, so can equally convincing and entirely compatible arguments be made about development's successes and failures. Any serious examination of the record of development will certainly document the impressive triumphs of **human development**: never before have the material circumstances and life opportunities of so many people been so drastically improved, in so little time. According to World Bank data, sustained growth over the past two decades has resulted in an 80 per cent increase in GDP per capita, despite substantial increases in population. Alongside income increases, remarkable progress has been made in literacy, education, life expectancy, food security, education, nutrition, and infant, child, and maternal mortality.

Those in the rapidly expanding global middle class enjoy standards of living that would not have been dreamed of by their grandparents. The pace of reduction in income poverty does not appear to be slowing; according to World Bank data the number of people living in income poverty worldwide has been reduced by almost half a billion in the space of just five years, from 2005 to 2010 (Chandy and Gertz, 2011), the fastest ever reduction of income poverty. The prime target of the Millennium Development Goals – to halve global income poverty by 2015 from its 1990 level – has thus already been achieved, well ahead of schedule.

Yet an assessment of the record of development must also recognize the escalating environmental and social costs associated with economic growth, and the fact that its benefits have been very unevenly shared both between and within countries. A large part of the progress that has been achieved can be attributed to massive poverty reduction in China alone. Beyond the small handful of rapidly emerging economies that have driven and benefited from global economic growth, large gaps still remain. Some 1.3 billion people do not have access to electricity, 900 million do not have access to clean water, 2.6 billion lack access to improved sanitation, and around 800 million rural dwellers do not have access to an all-weather road and are cut off from the world in the rainy season (Fay *et al.*, 2010; IEA, 2011). Inequality also remains significant in rich countries, with people at the bottom of the income ladder being excluded from many of the benefits of economic growth: at the time of writing one in six Americans, and a quarter of all British children, live in poverty (US Census Bureau, 2012; Save the Children, 2011).

As described in Chapter 7, the progress that has been achieved will be hard to replicate, let alone sustain unless human advancement can be decoupled from resource consumption and environmental impact. At a global level environmental costs are mounting and some planetary boundaries are already being breached (Rockström *et al.*, 2009). At the national level, environmental damages in many developing countries are reaching a point at which future development prospects are being undermined. According to a study of the World Bank, the damage done by environmental degradation is equivalent to 8 per cent of GDP across a sample of countries representing 40 per cent of the developing world's population (World Bank, 2011). In the words of noted ecological economist Herman Daly, economic growth is no longer economic: its costs are starting to outweigh its benefits (Daly, 1999). In other words, economic growth is failing in its own terms.

Economic growth of the kind experienced over the past two decades has been inefficient in terms of improving well-being overall: its inefficiency stems from the extreme concentration of capital in the hands of the already rich, with very little reaching those at the bottom of the income pyramid. As we have seen in Chapter 2, the well-being of the richest is hardly, if at all, affected by further income (see also Wilkinson and Pickett, 2010), whereas even small increases in the incomes of poor people can greatly improve their well-being. Even in China, where economic growth has lifted so many out of poverty, the country's leadership recognizes that current growth patterns cannot be sustained and that a change of course is needed towards an 'ecological civilization'. A central theme of the past two governmental Five Year Plans is the need to 'rebalance' the economy, moderate the pace of economic growth and improve its *quality*.

We are therefore faced with a big dilemma: conventional economic growth has been one of our most effective and time-tested means of achieving poverty reduction, yet it threatens to undermine the very social and environmental foundations for our future prosperity.

To conclude that there is a necessary trade-off between economic progress and environmental integrity would, however, be wrong. Rather, this dilemma reflects the narrow framing of economic goals and policies, and the overarching priority given to GDP growth in almost all countries. As a consequence the economy has been managed as if it were largely delinked from its social and environmental contexts. This delinked, market-dominated economy gives rise to a host of 'meta-externalities', in other words, the unintended and undesired side-effects of the economic system as a whole on the environment and on society. Global warming is one such meta-externality: a prime example of how economic growth has affected its own long-term sustainability (Stern, 2007).

Linear approaches to development are premised on emulating the experiences of rich and rapidly emerging economies (Willis, 2005). As outlined in Chapter 7, however, the GDP-centric pathways of development followed by roughly half a billion people in rich countries after the Second World War – built on abundant and cheap fossil fuels, large expanses of unexploited productive land and plentiful natural resources – are no longer available to poor countries today. There are clear signs that the climate and resource crunches are already upon us, and poor people are invariably the first to suffer. There simply are not enough of Earth's resources to satisfy the needs and appetites of a global, mass-consuming middle class projected to reach 5 billion by 2030 (from 2 billion in 2012), on existing patterns of consumption and production. The challenges of peak oil, recurring food and fuel price shocks, looming resource scarcities and climate change, are redefining options for human development in the twenty-first century.

While the processes of industrialization and the green revolution have delivered impressively on the promise of freeing humanity from material want and environmental constraints in the past century, the development challenge for the twenty-first century is to reground the economy in its physical and social contexts, so that it can continue to generate conditions for shared prosperity and well-being into the future. The sustainability challenge, illustrated in the Figure in Box 8.1, is to decouple improvements in human development from growing ecological impact.

Crucially, this dilemma of development in the twenty-first century is not confined to developing countries – its scope is universal. Within a shrinking 'ecological space' a fair sharing of finite resources, and poor countries' development prospects, depends on rich countries making room for them to grow.

Box 8.1

The historical relationship between human development and ecological footprint (1970–2010)

The twenty-first-century sustainability challenge is depicted in the figure overleaf, which shows the relationship between ecological impact and human development over time. The lines in the graph are trend lines for individual countries from 1970 to 2010. The data point for year 2010 is marked by the dot. The horizontal line denotes the world bio-capacity available per person (which shifts downwards as the global population increases) and the vertical line marks the threshold of high human development (>0.8 HDI). The bottom right shaded area represents the 'sustainability quadrant', in which the HDI reaches an acceptably high value but the ecological footprint remains with the limits of one planet Earth. Currently, no country achieves these two levels simultaneously. However, a promising development, shown by the downwards-sloping trajectories of some countries, is that they have improved well-being while reducing both natural resource demand and pollution.

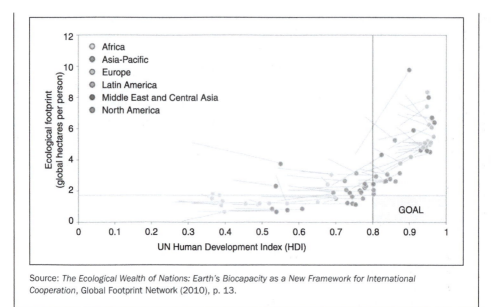

Source: *The Ecological Wealth of Nations: Earth's Biocapacity as a New Framework for International Cooperation*, Global Footprint Network (2010), p. 13.

2. Alternative development paradigms

Standard GDP-centric and resource-intensive approaches to development are at odds with the ethical imperatives of **equity** and sustainability. This conflict is rooted in the realm of ideas, in a particular paradigm of social change that posits the economy as detached from its social and environmental foundations. The result is a conflation of means (economic growth) with ends (human development), and a narrow framing of economic goals and policies that prioritizes the private over the public realm and externalizes social and environmental costs associated with the actions of private economic agents.

The famed Nobel economist, Milton Friedman, appreciated the importance of ideas as well as the way in which crises bring them to the fore. In his words:

> Only a crisis – actual or perceived – produces real change. When that crisis occurs, the actions that are taken depend on the ideas that are lying around. That, I believe, is our basic function: to develop alternatives to existing policies, to keep them alive and available until the politically impossible becomes the politically inevitable.

(Friedman 1962: xiv)

Box 8.2

Worldviews and alternative paradigms

In their examination of the relationship between globalization and the environment Clapp and Dauvergne (2005) identify four major normative 'worldviews' on environmental issues. Linking these to the climate change and development nexus can aid understanding of the complexity of different approaches and perspectives (see Hiraldo and Tanner (2011) for an example of the ideological drivers of REDD+). The four categories include:

● **Market liberalism**: Underpinned by the idea that economic growth and high per capita incomes are essential for human welfare and the maintenance of sustainable development. Stresses the role of the private sector and the use of market-based instruments in tackling climate change.

● **Institutionalism**: Centers on the need for strong institutions, good governance and effective laws to protect the environment and human well-being. Focuses on the regulatory regimes that govern climate change actions.

● **Bio-environmentalism**: Characterized by ecological limits and the need to modify human behavior in order to solve global environmental problems. Opposed to the business-as-usual model and driving ambitious targets for reductions in emissions.

● **'Social green' thinking**: Believes environment and society cannot be regarded as separate issues. Climate change responses must therefore balance emissions reductions goals with the well-being of human populations, stressing participation, rights and knowledge.

This section considers alternative development paradigms that propose solutions for marrying and advancing economic, environmental and social goals. It is beyond the scope of this book to cover them all. Rather, main alternative paradigms are highlighted as a means to illustrate how different ideologies or worldviews provide a normative basis to the relationship between environment and development and give rise to different perspectives and prescriptions for tackling the climate change challenge (see Box 8.2).

2.1 People-centred paradigms

Orthodox development approaches have been critiqued for prioritizing economic growth above all else and for their reductionist focus on material aspects of well-being (Willis, 2005). By contrast, a central tenet of people-centered approaches to development is the notion of individual sovereignty, of people as active agents in the process of societal change (Chambers, 1983; UNDP, 1990; Sen, 1999; Korten, 1990). A chief concern is to protect and advance human rights both as a means and an end of development. The technocratic, mechanistic and top-down view of social change typified in conventional trickle-down economics and standard development approaches – a view of development that fits well with the operational imperatives of large bureaucracies – is eschewed in favor of a view of development as a primarily political process, aimed at transforming power relations in society.

People-centered approaches to development generally view economic growth as necessary but by no means sufficient for improving well-being and eradicating poverty. They adopt a multi-dimensional view of poverty and emphasize the relative and subjective nature of well-being. Thus these approaches are inherently attuned to concerns of equity and **social justice**. They reject the view of poverty and inequality as residual outcomes of economic growth that can be addressed through discrete and targeted supplemental policy interventions, and instead promote integrated approaches to advancing social and economic goals. They go beyond material aspects and stress the importance of political freedoms, transparency and accountability in the actions of businesses and governments, and public participation in decision-making.

Another core principle of people-centered approaches is the emphasis on local control and ownership of local resources and of the development process, spurring greater self-reliance, accountability and creating incentives for more responsible stewardship of local resources that is essential to sustainability. Thus, sustainability is an inherent component and explicit goal of people-centered development.

The concept of people-centered development gained prominence in the 1990s, following the 'lost development decade' of the 1980s that had witnessed large-scale development reversals. It was advanced in several international development conferences in the 1990s, such as the Earth Summit in 1992, the International Conference on Population and Development (ICPD) in 1994, and the Summit for Social Development of 1995.

More radical variants of people-centered development, as put forward in the work of Robert Chambers (1983) and David Korten (1990), call for an upending of relationships of authority and a subversion of power structures, with an emphasis on empowering local actors to drive the change they want to see. Korten, for example, calls for fundamental transformations of our institutions, technology, values, and behavior 'consistent with our ecological and social realities' (Korten, 1990). These approaches are stridently opposed to economic liberalization, privatization and deregulation, and to market-driven economic globalization.

These extreme versions of people-centered development do not offer convincing responses to the climate challenge: being wholly focused on the local realm and decentralized, bottom-up decision-making, they are blind to the need for international collective action to address global challenges.

Mainstream approaches to people-centered development are represented by the 'human development' (Shahani and Deneulin, 2009) and 'rights-based' approaches (Mitlin and Hickey, 2006), the principles of which have been widely embraced by international development organizations. The human development concept has been widely promoted by the United Nations Development Programme's (UNDP) Human Development Reports since 1990. These approaches typically advocate for a set of core universal principles, such as human rights, freedom of speech, accountability, or rights to education and health care. This is in contrast to the more contextual or 'relativist' approaches put forward by Korten and Chambers that stress the differences in development between different societies and contexts.

Human development and rights-based approaches are primarily focused on social sector investments and achievements. Economic growth is seen as an important contributor to well-being that nevertheless needs to be guided by social goals and supplemented with proactive investments in **public goods** and social sectors, with priority usually given to addressing gender imbalances. Ultimately these approaches call for the integration of social and economic policies, and for the empowerment of peoples as agents in charting new, planet-sensitive development pathways.

2.2 New economic paradigms

Another influential set of approaches start from the premise that to meet the sustainability challenge requires re-wiring our economies. They rightly recognize that humans' impacts on the climate and environment are mediated through the economy. These approaches are steeped in a common view of the environment as natural capital constituting a crucial component of our wealth and well-being. They regard efforts to protect and enhance the environment as investments rather than sunk costs and they share a concern to better align economic incentives with ecological imperatives.

Recent calls for a new economic paradigm sprung from widespread dissatisfaction with conventional economics in the wake of the global financial and economic crisis, which was seen as symptomatic of deep structural flaws in the global economic system arising from the dominance of economic theories built on flawed reductionist assumptions about human motivations and behavior (Stiglitz, 2010; Kahneman, 2011). These assumptions have been systematically picked apart and critiqued by behavioral economists such as Princeton University's Daniel Kahneman, who offer more sophisticated and science-based alternative models of human behavior on which improved economic theories and policies can be constructed (see Table 8.1).

Table 8.1 *Evolutionary economics compared to neoclassical economics*

Issue	Complexity and evolutionary economics	Traditional economics
Micro-Foundation	Bounded rational agents learn and adapt their behavior; knowledge comes in different forms (e.g., tacit, codified) and is different from information; agents do not have complete information	Perfect rational agents make no errors, do not learn and have complete (or almost-complete) information
Dynamics	Open, non-linear systems usually out of equilibrium	Closed systems, static, linear systems in equilibrium
Links Among Agents	Many different types of links (technological, financial, personal, regulatory), including co-operative and competitive	Links among agents occur through market mechanisms
Relation Between Micro- and Macroeconomics	No division between micro- and macroeconomics; macro-patterns emerge from micro-level behavior and links	Micro- and macroeconomics are different disciplines
Evolution	Evolutionary processes of variation, selection, and retention at many levels (technologies, firms, industries) provide novelty	No explicit mechanism processes, technologies and product emerge (no mechanism for novelty)
Theory Development	Basically inductive: systems are discovered, as in natural science	Basically deductive
Policy	Specific to national, regional and sectoral systems and their specific context	One-fits-all policy prescriptions
Models	Complexity models, game theory, statistics, scientometrics	Calculus, algebra
Economic Growth	Occurs through creation of new products, and sectors	Occurs through addition of capital and labor and increase of productivity
Technology	Studied as endogenous to economic system and key determinant of growth	Exogenous to the economic system
Time	Timescales are key (i.e., for learning, institutional change)	Time is usually out of the models

Source: Tawney et al., 2011. Based on the work of Niosi, 2010.

Green growth

'Green growth' sits naturally within the standard intellectual framework of classical economics, upholds efficiency as a core value, and favors market-based solutions. It is no surprise therefore that the green growth approach has been most vigorously promoted by international organizations such as the World Bank and OECD.

Green growth reforms are mainly aimed at correcting **market failures** and 'getting the prices right' by introducing environmental/carbon taxation and reducing inappropriate subsidies to correct for externalities, creating tradable property rights, filling information gaps between economic agents, and investing in public goods (World Bank, 2012). Green growth policies contribute to economic growth in four ways (see graphical representation in Figure 8.1 below):

(a) increasing the flows of capital that can be brought into productive activity (represented in arrow (i) in Figure 8.1) – natural capital is boosted by sound environmental management; labor inputs are increased through reduced exposure to environmentally related health problems; and increased **resilience** to natural **hazards** means that physical capital is better protected;

Figure 8.1 *Analytical framework for Green Growth.*

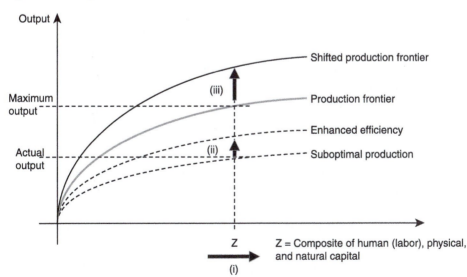

Source: World Bank, 2012c.

(b) improving efficiency (represented in arrow (ii) in Figure 8.1) – enhanced resource-use efficiency leads to economic savings, especially in an economy where prices fully reflect environmental costs and benefits;

(c) stimulating the economy, when it is functioning below peak capacity – as in a recession when unemployment and low demand create redundancy in the economy;

(d) spurring innovation (represented in arrow (iii) in Figure 8.1) – internalization of market externalities and policies to support R&D spur greater investment in clean technologies, resource-use efficiency and renewable energy innovations, resulting in an expansion of the production frontier.

Jacobs (2013) argues that the 'standard' version of green growth asserts the long-run economic benefit of environmental protection, while a 'strong' interpretation claims that environmental policy can in fact be a driver for growth. For both, investment in the environment is a means of protecting or inducing further growth, rather than for its own sake. The 'green growth' approach has been received with some skepticism in developing countries, particularly among fast-growing emerging economies, as it is seen as putting most of the burden of reform on developing countries, which will be the main sources of economic growth globally.

Green economy

The terms 'green growth' and 'green economy' are often used interchangeably, and while they do indeed point to many similar policy prescriptions, they reflect quite different philosophies both about the end points of social change and about the respective roles of the state and markets in bringing about desired changes. In terms of policy prescriptions, the 'green economy' can be said to encompass 'green growth' but goes beyond it in ambition and scope. Whereas 'green growth' is about greening the increment of economic activity, 'green economy' approaches are also about correcting existing imbalances in the allocation of capital throughout the economy. The green economy is therefore more prescriptive about what the economy should deliver in terms of social and environmental outcomes.

The United Nations Environment Program (UNEP) has been a vocal proponent, defining a 'green economy' as 'one that results in improved human well-being and social equity, while significantly reducing environmental risks and ecological scarcities. In its simplest expression, a green economy can be thought of as one which is low carbon, resource efficient and socially inclusive' (UNEP, 2012: 1). In this approach, the government is a much more active and pervasive force in directing the economic activity and resource allocation – e.g. through direct government spending and price supports. This is in contrast to green growth approaches, which put more faith in the market as the main and most efficient resource allocation mechanism. According to UNEP, a green economy entails proactive mustering of the economic resources by the government to 'maintain, enhance and, where necessary, rebuild natural capital as a critical economic asset and as a source of public benefits, especially for poor people whose livelihoods and security depend on nature' (UNEP, 2012: 1). The agenda therefore carries with it an implicit critique of free-market economics, whereas green growth approaches constitute an improvement and extension of the free-market paradigm.

Steady-state economy

In more radical conceptions of the green economy, the possibility of boundless economic growth is rejected, and a 'steady state economy' (SSE) – that is, a no-growth economy characterized by equilibrium conditions – is seen as the only truly green economy. The fact that our economies are geared only for expansion is seen as the root of our environmental and societal problems and humanity must kick the growth habit if it is to survive and prosper (Jackson, 2011). The implication is a fundamental re-wiring of our economies so that growth is no longer its modus operandi.

Steady-state economics is in the intellectual tradition of the Club of Rome's *Limits to Growth* (Meadows *et al.*, 1972), popularized by authors such as Herman Daly (1996) and Tim Jackson (2009). A central tenet of SSE is that size matters. It argues that as the economy grows and reaches a planetary scale (a scale at which it affects and interacts with global biophysical processes), it must eventually conform to the behavioral principles or 'laws' that govern biophysical systems. As resources start to run out, maintenance of the resource base becomes a paramount economic management imperative. Thus a central feature of the steady-state economy is a constant stock of physical capital that can be maintained by a low rate of energy and material resource use ('throughput') that lies within the regenerative and assimilative capacities of the ecosystem (Daly, 2005).

A key concern for proponents of the SSE is to stabilize the economy at a size that is compatible with the maintenance of key ecological functions. The four commonsense rules of a SSE are (Daly, 2005):

- Maintain the health of ecosystems and the life-support services they provide.
- Extract renewable resources (e.g. timber, fish, etc.) at a rate no faster than they can be regenerated.
- Consume non-renewable resources (e.g. fossil fuels, minerals, etc.) at a rate no faster than they can be replaced by the discovery of renewable alternatives.
- Deposit wastes in the environment at a rate no faster than they can be safely assimilated.

An end to economic growth as we know it need not mean an end to development. Indeed, SSE proponents are adamant that the two are not the same. In the words of the ecological economist, Herman Daly: 'growth is the quantitative increase in physical scale while development is qualitative improvement or the unfolding of potentiality. An economy can grow without developing, or develop without growing, or do both, or neither' (Daly, 1996). Even the early political economist John Stuart Mill did not envisage economic activity expanding ad infinitum – in his words:

> . . . a stationary condition of capital and population implies no stationary state of human improvement. There would be as much scope as ever for all kinds of mental culture, and moral and social progress; as much room for improving the art of living, and much more likelihood of it being improved, when minds ceased to be engrossed by the art of getting on.
>
> (Mill, 1848: Book 4, Chapter 6)

2.3 New environmentalist approaches: Towards planetary management

Another variation on the notion of development as an environmentally constrained process stems from the planetary boundaries literature, complexity theory and the Anthropocene theory. The challenge of development is cast in terms of epochs, or ages: the new rules of the Anthropocene age require that we actively engage in planetary management to ensure we stay within the 'safe operating space' set by global environmental thresholds (Steffen *et al.*, 2011). Whereas environmentalism traditionally focused on reducing human interference with nature, the Anthropocene calls for a different kind of environmental stewardship since avoiding human impacts on the environment is no longer an option. The environmental challenge then is to actively *manage* rather than reduce human impact, i.e. to choose what aspects to minimize, tolerate and adapt to. Rather than being 'protectors' of the integrity of the natural world, we need to be intelligent designers at the planetary level in order to help Earth regain its balance (Folke *et al.*, 2011; Lynas, 2011).

The focus is not on people, or on economic systems, but on human ingenuity, on using the powers of science and technological innovation to manage the planetary systems to our advantage, so that our species can continue to survive and thrive in a 'full' world shaped by our impact. Powerful technological options like nuclear, synthetic biology and genetic engineering need to be actively explored and developed rather than foreclosed (Lynas, 2011). In the planetary boundaries perspective, measures to buy time are important, and geo-engineering solutions could therefore play an important part in the planetary management toolkit.

Geo-engineering involves the large-scale manipulation of Earth processes for the purpose of counteracting climate change. Examples include carbon capture and storage (CCS) from coal-fired power plants, fertilizing oceans with iron to encourage the growth of CO_2-absorbing plankton, installing CO_2 scrubbers to remove CO_2 from the atmosphere, pumping aerosols into the stratosphere to act as a sunscreen, or deploying thousands of mirrors in space to deflect sunlight.

Some geo-engineering solutions are fanciful, and many are still years away from being viable at a large scale. More importantly, tampering with Earth processes can carry serious risks of adverse unintended consequences. Nevertheless, as the window for effective climate action narrows, geo-engineering solutions are increasingly emphasized as an important component of the climate policy toolkit (see Chapter 5 of Pielke (2010) for an excellent discussion of the scope and limitations of geo-engineering solutions).

Whereas people-centered approaches focus on the local level, the planetary boundaries analysis stresses the need for collective action to be taken also at the *global* level. A strong emphasis is placed therefore on the global governance structures and international cooperation, which are needed to manage the kind of epochal decisions that will need to be taken, such as implementation of geo-engineering responses. In order to effectively shift from resource exploitation towards stewardship and management of the Earth system, another important feature of decision-making and institutions is that these will need to encourage **experimentation** and allow for systematic **learning**, as we would be operating in uncharted waters.

The various approaches/paradigms are presented above in slightly caricatured form to emphasize their differences and highlight characteristics. They are not all mutually incompatible, and some approaches draw from several of these paradigms. For example, Oxfam senior researcher Kate Raworth has combined insights and methods from both the people-centered and the planetary boundaries approaches to define a 'safe and just space' within which humanity can thrive in the twenty-first century. That space, shaped like a doughnut, is delineated by a social foundation (inner boundary) as well as environmental thresholds (outer boundary) (see Figure 8.2). The social foundation refers to the achievement of enabling conditions for human development, while the environmental threshold is defined by the planetary boundaries. In her words, 'Moving into this space demands far greater equity – within and between countries – in the use of natural resources, and far greater efficiency in transforming those resources to meet human needs' (Raworth, 2012: 1).

3. Building blocks for an inclusive and sustainable future

The different approaches and paradigms are presented above to emphasize their differences and highlight key characteristics. Nevertheless, there is considerable overlap in terms of the practical suggestions of steps to be taken. They often agree on the need to build the enabling conditions for the transition away from GDP-centric and resource-intensive modes of development towards more sustainable and inclusive development pathways. The common ingredients, or building blocks, of this transition are presented below.

Figure 8.2 *Living within the Doughnut: 'A safe and just space for humanity.'*

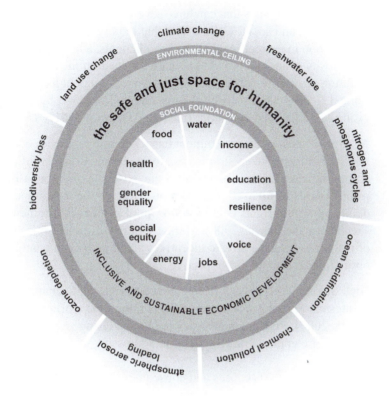

Source: Raworth, 2012, Credit: Kate Raworth.

3.1 Reframing the government's role in economic and environmental management

The need for smart governmental intervention to address climate change and related challenges constitutes a common denominator of all the approaches discussed above. It is worth restating the rationale for public action, because there is a stubborn strand of public opinion that holds that – for climate change as for any societal ill – government is the problem rather than the solution, and that if only government were to step out of the way, the free choices and interactions of rational agents (individuals, companies) would collectively deliver the best possible outcome socially. Even in the world of textbook economics, though, free and competitive markets only deliver the most efficient and socially optimum outcome under certain technical assumptions. Real markets deviate from the 'ideal' in a multitude of ways that can have severe consequences for the environment and social welfare, and that require government intervention to correct. As Tim Jackson laments, a large part of the reason why economic growth has started to undermine the very basis of future prosperity is because 'the role of government has been framed so narrowly by material aims, and hollowed out by a misguided vision of unbounded consumer freedoms' (Jackson, 2011: 15).

At a most basic level, government has a key role to play in correcting market failures. The economic case for government intervention to correct **market failures** arises from the existence of **externalities, public goods, informational asymmetries** and the lack of clearly defined property rights over **common pool resources** (Sterner, 2003; Tietenberg and Lewis, 2000). Market failures result in a misallocation of resources because private incentives are at odds with what is optimal from a societal perspective.

Nick Stern and his review team characterized climate change as the greatest and most wide-ranging market failure the world has ever seen (Stern, 2007). Five characteristics of climate change explain why a laissez-faire approach is not suited to addressing it:

- It is global in its causes and consequences: in the absence of a global carbon price, regulatory arbitrage by economic actors leads to carbon leakage.
- It occurs over long time scales, whereas markets are inherently myopic: the standard use of discounting (e.g. interest rates) in economic transactions and decision-making introduces a systematic bias against the future, and rationalizes inaction on issues that entail long-term costs or risks.
- It involves profound uncertainties and non-linear, non-marginal change such that individual actors cannot fully foresee the impacts of their actions let alone internalize these in their decisions.
- Impacts are cumulative; addressing climate impacts therefore effectively requires significant interventions long before the most severe consequences of the problem manifest.
- There is a degree of irreversibility that results from the inertia of climate systems.

Stern prescribes three kinds of governmental intervention to effectively respond to climate change. The first is the **pricing of carbon,** implemented through carbon taxation, trading or regulation. The second is action to **remove barriers to energy efficiency** (e.g. through performance standards and building codes) and to inform, educate and persuade individuals about what they can do to respond to climate change. And the third is policy to **support innovation and the deployment of low-carbon technologies** (Stern, 2007). The government can play a useful role in that regard by serving as a 'proto-market' for new technologies (e.g. as demonstrated in government's role in developing the internet and satellite communications).

The scale of reductions needed to avert catastrophic scenarios is well beyond historically observed rates of 'decarbonization', that is the reductions delivered through the autonomous actions of individuals and the market. Government intervention is necessary to nudge private actors and markets beyond incremental improvements in resource- and carbon-efficiency towards embracing transformative shifts. To give private actors the confidence to invest in the risky business of disruptive change requires sending long-term policy signals through bold government commitments to the three kinds of market interventions described above.

Putting market failure aside, there are compelling technical and ethical reasons for governmental intervention. The potential to reap development co-benefits – e.g. reduced air pollution from emissions reductions – and the presence of significant barriers to autonomous **mitigation** and **adaptation** constitute strong technical arguments for government policy support (see Chapter 4). As examined in Chapters 4, 5 and 6, barriers can be financial (cost barriers, imperfect markets, etc.), informational, institutional and political.

The need to address issues of equity and justice also constitutes a strong ethical argument in favor of government intervention: those with the strongest stake in effective climate policy – the poor, the young and the generations to come who will be most hurt by the impacts of climate change – lack power and voice to shape an effective response (Roberts and Parks 2007; see Chapter 3). Unfettered markets are skewed by the concerns and needs of affluent people. Thus autonomous responses to climate risks can in themselves impose costs on poor people, as when flood protection measures by the rich result in floodwaters being diverted towards areas inhabited by poorer people.

Social policy (e.g. safety net and cash transfer programmes) can play an important role in building resilience among poor communities to the impacts of climate change and reducing the costs of transitioning towards climate-smart development pathways, and as such will be an important feature of any policy toolkit to address development impacts of climate change (Davies *et al.*, 2009).

Importantly, however, *more* government intervention is not in itself always a good thing. Government action is also subject to failure, sometimes proving an obstacle to effective climate action. A case in point is the use of perverse subsidies for fossil fuels – these exceeded US$650 billion in 2008 – which undermines efforts to transition towards renewable energy solutions (IEA, 2011).

Rather than more government, it is *smarter* government that is needed. Governments need to become much better at tackling uncertainty and long time horizons and at integrating concerns of all citizens in development and climate decisions (WRI, 2011). Some decisions call for a stepwise approach that keeps future options open and avoids '**lock-in**' to future **vulnerability**. Other decisions, however, with long-term consequences, call for early choices to take aggressive action with future risks in mind. Decision makers must learn to take threshold effects into account, and be flexible to allow for mid-route changes of course as new information becomes available and learning takes place (more on all this in section 3.4 of Chapter 5). Greater transparency and accountability will be crucial to making governments smarter and more responsive to the needs and concerns of its citizens, rich and poor.

The pervasive nature of the climate challenge means that it cannot be tackled in isolation from broader questions of economic management (e.g. trade, debt, fiscal policy) and sectoral planning. The case for policy coherence and for integrated approaches to climate and development policy making was presented in Chapter 2–5. Integrated responses may reap important benefits and are necessary to realize synergies and manage trade-offs.

Policy integration requires institutional strengthening and innovation to support adaptive learning and cross-governmental coordination. One example of this is the establishment in Kenya of a Climate Change Coordination Unit within the Office of the Prime Minister to oversee and coordinate climate policy formulation and implementation across government. It also requires the generation of decision-relevant information. There will be a general need for more holistic knowledge, cross-disciplinary research and systems analyses (Grist, 2008; Folke *et al.*, 2011). As important as creating that knowledge is ensuring that decision makers – from small-scale farmers, to city mayors, to national-level ministers – are able to easily access the information they need to manage climate risks (Hellmuth *et al.*, 2007).

3.2 Strengthening international collective action: Linking climate and development

Chapters 2 and 3 set out the need for global collective action. A strong and effective international climate **regime** provides a crucial underpinning for such action. Yet while there have been important achievements in building the climate regime over the past 20 years, progress to date has fallen far short of what is needed. Countries have largely failed to live up to their promises in terms of emissions abatement and financing pledged to help poor countries join in the global effort (Roberts and Parks, 2007). Even if promises were kept, the level of ambition in current pledges would still be far off the mark, resulting by 2020 in emissions levels 27 per cent higher than those required by the 2°C **stabilization** target (UNEP, 2010).

For many, the low point of international climate action came at the COP15 **UNFCCC** meeting Copenhagen in December 2009. Few were satisfied with the last minute 'accord' that was struck behind closed doors by five of the largest nations in the world, after two weeks of discussions had failed to deliver a consensus agreement. However, this event signalled an important turning point in the evolution of the international climate regime. The pivotal role played by the BASIC (Brazil, South Africa, India and China) group, the loosening of traditional negotiating blocs and subsequent formation of new alliances straddling both richest and poor countries heralded a new age of international climate politics, defined less by the North–South divide and reflective rather of a multi-polar world.

The Copenhagen climate conference at once demonstrated the necessity and the insufficiency of the UN system as a fulcrum of international climate cooperation: only the UN could provide an effective platform for the smallest and poorest countries to voice their concerns and have their positions seriously considered, yet inclusiveness in decision-making can also lead to delay, stalemate and un-ambitious compromises.

The Copenhagen conference illustrated the need for strengthening the multi-lateral UNFCCC process whilst pursuing complementary bilateral and 'plurilateral' strategies and transnational multi-stakeholder partnerships in parallel (Moncel *et al.*, 2011; Andonova *et al.*, 2009). Future historians may well look back on it as a turning point in international governance marked by a shift away from centralized and

Figure 8.3 *The importance of mitigation commitments from major emitters.*

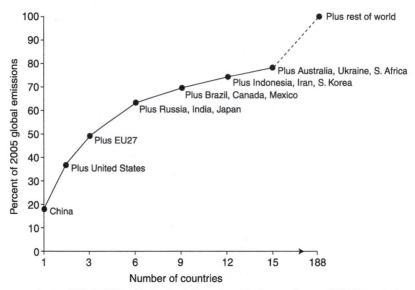

Sources and notes: WRI, CAIT (http://cait.wri.org). Percent contributions are for year 2005 GHG emissions only. Moving from left to right, countries are added in order of their absolute emissions, with the largest being added first. Figures exclude emissions from land-use change and forestry, and bunker fuels. Adapted from Figure 2.3 in Baumert *et al.* (2005).

hierarchical inter-governmental decision-making towards polycentric, multi-stakeholder and multi-level forms of governance to address global challenges.

The bilateral/plurilateral approach, characterized by alliances made up of two or more countries or regions, involves a limited number of countries working together around shared interests or sectoral agreements. Such approaches can be particularly important to striking agreements among major emitters or to foster innovation (Moncel *et al.*, 2011). Figure 8.3 illustrates the importance of such approaches: 14 of the biggest emitting countries plus the EU alone account for close to four-fifths of global emissions. Sometimes such agreements can be piggy-backed on to existing plurilateral processes, such as the Major Economies Forum, the G20 or the Association of Southeast Asian Nations (ASEAN). Plurilateral approaches have a number of potential benefits (Moncel *et al.*, 2011) including:

● More straightforward negotiation and implementation processes, with smaller numbers of countries coalescing around points of agreement and shared interest.

- Providing a platform for testing out innovative approaches that might be more difficult to agree in a multi-lateral process.
- Facilitating South–South cooperation, which is increasingly an important means for sharing adaptation/mitigation solutions and capacities between developing countries.
- Establishing momentum for increasing ambition and generating trust for the wider multi-lateral process.

Transnational multi-stakeholder partnerships such the World Business Council on Sustainable Development, or issue-specific 'clubs' such as the World Mayor's Council on Climate Change, C40 Cities Climate Leadership Group, will also have an important role to play in advancing international climate action.

As for the UNFCCC process, restoring trust among the different parties, especially between developing countries and richer OECD countries, will be crucial to enhancing its effectiveness. Roberts and Parks (2007) argue that this trust has been systematically eroded over the past two decades due to the consistent failure of rich countries to meet their promises both in terms of emissions abatement and pledged financing to assist developing countries in addressing the impacts of climate change and shifting towards **low-carbon development** pathways. One important aspect of trust-building is to develop and agree the means of monitoring, reporting and verifying (MRV) climate finance, actions and performance and progress against pledges (more on this in Chapter 6). Developing common standards of measurement will therefore be vital to giving more credibility to commitments.

On a deeper level, breaking through the cycle of mistrust requires tackling the inequalities that are deep-rooted in the international economic regime and perpetuated through a host of international institutions and treaties. Rich countries will need to demonstrate convincingly that they care about the well-being of people in poor countries and genuinely want to help them escape poverty and structural vulnerability and develop in a sustainable way. Development aid obviously has an important place in this but is by no means sufficient: reforming agricultural and international trade policies will have a far bigger positive impact.

In the current international climate regime, rich countries make demands on poorer countries through multi-lateral and bilateral agreements that foreclose the very development paths that they had themselves followed, what Cambridge economist Ha Joon Chang calls 'kicking away the development ladder' (Chang, 2002). The international treaties, aid conditionalities and international trade agreements they have thrust upon the developing world have the effect of restricting 'policy space' – the leeway for exploring diverse policy options – in developing countries. Robert Wade describes this pattern as 'the shrinking of development space' (Wade, 2003). These patterns are occurring at a time when policy experimentation is arguably more needed than ever before to chart new pathways to climate-smart development.

The upshot of the above is that restoring trust requires a new grand bargain between rich and developing countries that places fairness and equity at its very centre. Rich countries can signal their commitment to creating a more just and fair international order through a series of confidence-building measures (Roberts and Parks, 2007). In doing so they must acknowledge developing countries' right to develop and agree to provide greater 'environmental space' for them, while at the same time committing to leading by example in making drastic cuts to their **greenhouse gas** emissions. Rich countries can also take important measures to level the international economic playing field, such as eliminating or reforming agricultural subsidies that depress international agricultural prices and create unfair hurdles for developing countries' agricultural exports. Giving more say to developing countries in international decisions and rules setting, e.g. by further increasing their representation in the governance structures of international financial institutions, would also lend greater credibility to a new development grand bargain (Roberts and Parks, 2007).

3.3 Unleashing the transformational power of businesses

Businesses have an enormous influence on human and economic development and are major sources of environmental impacts around the world. Businesses stimulate economic growth, provide goods and services to meet the ever-changing needs of society – including the most basic of needs – create jobs and wealth in communities, and produce technologies that enhance quality of life and the productivity of the economy. They can play crucial roles as fountains of innovation and shapers of green markets and supply chains. But they are also chief culprits in driving unsustainable patterns of production and consumption. The trend over the past decades has been towards a growing influence

Figure 8.4 *Sources of Aid Funds from DAC Countries.*

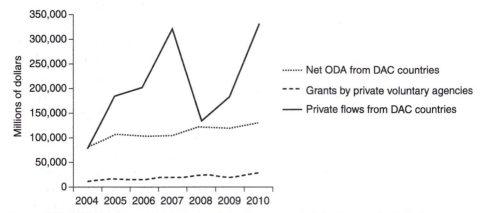

Net ODA from DAC countries

Grants by private voluntary agencies

Private flows from DAC countries

Source: OECD, 2012 'Development aid: Grants by private voluntary agencies', *Development: Key Tables from OECD*, No. 3. doi: 10.1787/aid-pvt-vol-table-2012-1-en.

of the private sector around the world. The private sector's contribution to North–South financial flows is also increasingly important, exceeding official flows for most of the past decade (see Figure 8.4).

As both a provider of environmental solutions and a source of today's climate and environmental problems, constructive engagement and partnership with the private sector will be crucial. Advancing sustainable development in the context of multiple accelerating pressures (e.g. resource scarcities, climate change) requires an economic transformation, a new 'industrial revolution' of the kind that can only be delivered in partnership with the private sector. The scale of investment, innovation, technology development and employment creation required is beyond the range of what the public sector alone can deliver. Governments will need to work closely with businesses to steer business investment and innovation toward advancing solutions to environment and development challenges. It should come as no surprise that many leading development agencies (including DFID, Danida, Sida, DGIS and the World Bank) have put private sector engagement at the center of their development strategies.

The challenges facing small- and medium-sized enterprises (SMEs) are quite distinct from those facing big companies; accordingly, unleashing their contribution to sustainable development requires different approaches. SMEs are often the engines of local economies in developing countries, where they contribute significantly to employment. They also account for a large share (60 per cent in many countries) of pollution and resource use, and therefore have a potentially pivotal role to play in improving and maintaining natural resources and the environment as well as in providing basic goods and services for poor people. However, SMEs face challenges in bringing their ideas to fruition and their products to market and most of them do not survive beyond the first few years of operation (Barreiro, Hussels and Richards, 2009). For SMEs that manufacture or supply environmentally friendly products and serve low-income communities the challenges can be particularly daunting.

One particularly debilitating obstacle facing SMEs is the lack of credit and financial resources to grow. According to the Global Partnership for Financial Inclusion, in 2011 almost three-quarters of all SMEs lacked access to credit, with difficulties particularly noted in Asia and Africa. In addition to difficulties accessing markets and attracting talent, SMEs tend to face daunting policy barriers that increase the cost of doing business and force many companies out of the market. The key to unlocking the positive potential of SMEs is often to improve the enabling environment and support system for small businesses to operate and thrive.

For larger companies, the challenge is to place climate, environmental and social issues at the core of business strategy and supply chain management. Companies such as Unilever, Nestlé and Wal-Mart and SAB Miller are some more famous examples of companies that have put sustainability at the forefront of their strategy. They did so in response to product and supply chain exposures to water and climate risks, dependencies on natural resources and other ecosystem services, and to get ahead of the curve in realizing business opportunities associated with the transition towards a new economy (Morgan, 2011; Hart and Prahalad, 2002). Governments will have an important role in building an enabling environment for businesses to profitably invest in tomorrow's green growth sectors.

3.4 Measuring what really counts – Beyond GDP

Measurement matters. It informs our perceptions of reality and hence our actions. It forms the basis of management: what cannot be counted cannot be managed. Better measurement is therefore key to making better decisions. In the words of Gus Speth, the founder and first president of the World Resources Institute: 'You get what you measure, so measure what you want.'

By far the most important and consequential measure in contemporary society is the economy's Gross Domestic Product (GDP). It represents in a single number the monetary value of the totality of a country's economic output. As such GDP provides an exceedingly important input to economic decision-making. Economic decision makers in the 1930s were effectively flying blind when responding to the onset of the Great Depression, and their bundled policies greatly exacerbated its devastating impact. To avoid a repeat of such disastrous economic policy making, Nobel economist Simon Kuznets was tasked to develop the GDP in 1934.

While GDP is a decent measure of economic activity transacted through the market, it is deeply flawed as a generalized measure of progress (see Box 8.3). Indeed, it was never designed for that purpose.

Box 8.3

The shortcomings of GDP

- GDP merely quantifies the monetary value of goods and services produced and exchanged. It says nothing about the contribution of these to human well-being.

- GDP sheds no light on sustainability. If current consumption were entirely financed by running up a debt that cannot be repaid, this doesn't show up in GDP figures.

- All non-monetary values and non-market transactions are excluded. The focus on what can be submitted to the measure of money leads to an overemphasis on formal markets, and insufficient attention to critical unpaid economic activities (e.g. parenting, volunteer work, etc.). Likewise, external (that is non-market) costs, e.g. of pollution and resource depletion are not reflected.

- GDP is blind to equity considerations as it says nothing about the distribution of wealth, income or opportunity within society. This leads to the paradox that countries like Swaziland and Angola have per capita GDPs of middle-income countries, but rank among the bottom achievers in terms of social and human development indicators.

- Most importantly, GDP fails to account for resource depletion or the degradation of the five kinds of capital on which our economy and well-being depends. This matters a great deal, for if the environment were properly treated as productive capital, it would make economic sense to invest in it, and environmental policies would be considered as investment, rather than imposing costs. Future growth may be compromised if alternative investments are insufficient, or if critical thresholds of natural capital are reached, undermining economically important or vital ecosystem services (UNEP, 2012).

Source: Daly, 1996.

It tells us very little about many of the issues that matter to many of us, such as environmental quality, income distribution, public goods and employment. Yet it has come to be interpreted universally as synonymous with welfare and social progress, and as the ultimate yardstick of a country's strength. As a result, it has skewed economic policy and political goals towards the single-minded pursuit of GDP growth.

From the very outset, Kuznets was aware of the risk that GDP would be misused: he cautioned in his first report to the US Congress on GDP in 1934 that 'the welfare of a nation can [. . .] scarcely be inferred from a measure of national income . . .'. In a 1962 paper, Kuznets reinforced and further elaborated on this point, criticizing the extent and scope to which GDP had been used: 'Distinctions must be kept in mind between quantity and quality of growth, between costs and returns, and between the short and long run. Goals for more growth should specify more growth of what and for what.' (Kuznets, 1962.)

Attempts to develop better measures of progress fall into three main categories. First are attempts to fix GDP, generally by broadening the scope of what is taken into account in its calculation, and distinguishing between benefits and costs, or more precisely between what adds to or subtracts from well-being. Second are attempts to construct alternative composite indicators. Third are attempts to go beyond a single measure and to promote instead a 'dashboard' of indicators that taken together give a fuller and more accurate picture of the true complexity of human progress.

Fixing GDP involves putting economic values on a range of things that have not been counted as part of the market economy and that matter for our well-being, such as housekeeping and childcare, leisure time, or ecosystem services. The aim is to have a more accurate and realistic gauge of the level and changes in total wealth, including stocks of physical, human and natural capital. Adjustments are made downward or upward to reflect negative or positive changes in welfare that can be monetized through standard valuation techniques. For example, while standard GDP measures attribute no value to natural resources until they are extracted and sold, adjusted GDP measures would register the depletion of natural resources as a cost that drags down GDP. Likewise, GDP would be revised downward to reflect the health costs associated with pollution. Conversely GDP would be adjusted upward to include the social benefits of higher education, and the benefits of sustaining ecosystems for current and future generations, etc.

A crucial element of any effort to better manage our economy's impact on the climate and environment is to enhance and value the economic benefits that the environment provides seemingly for free, i.e. 'to make the work of the biosphere visible in society, in human actions and in financial and economic transactions' (Folke *et al.*, 2011). Factoring in the full social costs of fossil fuel use will also encourage a shift to cleaner fuels. The World Bank's Genuine Savings and Natural Capital Accounting initiative, Green GDP, Genuine Progress Indicators, and the System of Environmental and Economic Accounting (SEEA) developed by the UN Statistical Division are some examples of adjusted GDP. These have been bolstered by a decision by the UN Statistical Commission to endorse international methodology for environmental accounting.

Composite indices go beyond money metrics and economic accounting as a means of indicating progress. Indices may be particularly helpful for performance benchmarking and comparative policy analysis. They combine a range of indicators that often use different units of measurement (e.g. income, life years, mortality rate, educational attainment, etc.) weighted and normalized to a value typically expressed on a scale between 0 and 1. These are often presented in the form of country rankings. An example of an index is the 'Environmental Sustainability Index' (ESI) developed by Dan Esty at Yale University (Esty, 2001). The ESI covers five components of sustainability: the health of environmental systems; anthropogenic stresses on the environment; human vulnerability to environmental stresses; societal capacity to respond to environmental challenges; and global stewardship (cooperation with other countries in the management of global environmental commons).

The most widely used composite index, constructed to challenge the dominance of GDP as a measure of progress, is the UN Human Development Index (HDI). The HDI is a global ranking that incorporates two other factors alongside GDP: educational attainment and citizens' health (see Box 8.4). It aims to capture explicitly those factors that matter most in terms of determining peoples' capabilities and opportunities in life.

More recently there has been increasing attention on the use of 'dashboards' or sets of indicators capturing several aspects of

---

Box 8.4

The Human Development Index (HDI)

The inadequacy of income for representing people's well-being has been the central message of the human development and capability approach. Informed by the *human development* thinking and as a rejection of using GDP as the sole measure of progress, the Human Development Index (HDI) was first introduced in 1990 with the objective of measuring human progress and quality of life at the global level. It has since become one of the most influential indicators to measure a country's development progress.

The HDI is a composite index that summarizes a country's total achievement in three dimensions of human development: health, education and income. It intends to reflect achievements in the most basic human capabilities: living a long life, being knowledgeable and enjoying a decent standard of living.

Given that the human development approach supposedly intends to measure objectives of human development (*ends*), it has been criticized for retaining income as one of the relevant dimensions, because in the human development approach income is regarded exclusively as a *means*. Its simplification of the breadth of the human development and capabilities approach into just three factors has also been questioned. One response has been to create HDI-hybrids that cover specific issues, including a Gender-related Development Index (GDI), the Gender-Empowerment Measure (GEM) and the Human Poverty Index (HPI).

Source: Adapted from Deneulin and Shahani, 2009.

---

development and its sustainability. In 2009, President Sarkozy of France created a Commission on the Measurement of Economic Performance and Social Progress, bringing together the world's greatest economic minds, chaired by Nobel economists Amartya Sen and Joseph Stiglitz, and French economist Jean-Paul Fitoussi. The task of the Commission was to investigate alternatives to GDP as a means of better measuring economic performance and social progress (Stiglitz, Sen and Fitoussi, 2009).

While recognizing the political appeal of a single headline measure to replace GDP, the Commission recommended instead that a dashboard of indicators be used to complement GDP. The metaphor of a car was used to explain this. Relying on GDP alone is tantamount to driving in a car with only one dial on the dashboard indicating speed: the driver has no way of knowing other critical information such as fuel levels, distance travelled, or engine temperature. At the same time, there should not be more indicators than can easily be digested by decision makers or the public; the dashboard should be big enough that the driver at the wheel can take in the information needed to drive safely. According to the Commission, a good dashboard for assessing well-being will require metrics from at least seven categories: health, education, environment, employment, material well-being, interpersonal connectedness and political engagement. Measuring equity, such as the distribution of material wealth and other social goods, is also important (Stiglitz, Sen and Fitoussi, 2009).

3.5 Achieving transformational change

Chapter 7 outlines the need to review our approach to development in light of climate change and broader shifts in the international landscape. As a result of the insufficient scale and speed of international climate action to date, the global temperature rise is now likely to exceed 2°C and require adaptation to more dangerous levels of change of 4°C or above (New et al., 2011; Stafford Smith et al., 2011; World Bank, 2012d). Averting this catastrophic scenario will require rapidly scaling up climate-smart solutions around the world. Catalyzing the emergence and rapid spread of new climate smart innovations will be key to effective development and climate action. The same is true for other planetary boundaries. A series of scenarios produced by the Tellus Institute using its Polestar modeling approach and planetary boundaries data indicate that even with far-reaching policy reforms implemented across nations and within key sectors, from energy supply to food production, the world is still likely to transgress critical planetary boundaries, in particular loss of biodiversity (Gerst and Raskin, 2011). With billions set to join the global middle class and growing concern about equity and 'fair shares' in an environmentally constrained planet, even a coordinated and sustained programme of incremental adjustments may be insufficient to avert sliding into a planetary danger zone.

Paradigms of linear causality condition much development thinking and practice. They give rise to command-and-control hierarchies and centralized decision-making that often dampen creativity and innovation. Yet 'coping with past problems often creates dysfunctional systems. To meet coming challenges requires transformations of world views, institutions, approaches, and methods' (Folke et al., 2011: 721). Increasingly, researchers and practitioners alike are accepting that meeting the twin climate and development challenges requires radical approaches that spur behavioral and technological change at a different speed and magnitude, fundamentally altering the existing links between economic and environment systems.

The term 'transformation' has been used in many ways, but is most commonly interpreted as radical change requiring innovation and testing of new approaches. Main characteristics of transformation are summarized in Box 8.5.

Box 8.5

Characteristics of transformation

A review of the literature on transformation and interviews with experts in the context of climate change and development revealed a set of common characteristics of transformational change that include:

1 **Radical change, innovation and experimentation**
   Transformation is commonly interpreted as radical change requiring innovation and testing of new approaches. This entails the generation/use of new knowledge and a markedly different way of doing things.

2 **Addressing power imbalances**
   Addressing power relations and helping individuals to be 'empowered' lies at the core of an intervention becoming transformational. Success involves recognizing the social and political processes underpinning vulnerabilities, building inclusive forms of governance, redistributing benefits and more equal access to resources.

3 **Critical reflection, beliefs and values**
   Transformation entails being aware of and challenging inherited ways of thinking, assumptions and biases, as well as recognizing and negotiating power structures. This involves developing alternatives to entrenched and institutionalized positions, reflecting critically on one's own patterns of behavior and addressing moral trade-offs in change processes.

4 **Effective leadership**
   There is a need for leadership that can perceive the need for transformation and communicate this effectively. Leaders can challenge the status quo, provide alternate visions of what is possible, take advantage of policy windows and manage conflicts that may emerge during transformational processes.

5 **Collective vision and future orientation**
   A collective vision for change is crucial in transformation, developed in advance but often through trial periods and experimentation. Preparing strategies in advance enables them to take advantage of crises as opportunities for their uptake. Transformational initiatives are therefore oriented around long-term visions of the future.

6   **Moments of opportunity and policy windows**
Without the ability to open or recognize policy windows, the aggregation of the
other components of transformation may not be employed usefully to drive change.
Understanding the sources of stimulus of policy windows is critical to processes of
transformation.

Source: Tanner and Bahadur, 2013.

Transformation as a concept draws on adaptive cycles in resilience
theory and transitions theory (van der Brugge and Rotmans, 2007;
Nelson *et al.*, 2007), which describes four phases of change that
include:

- early experimentation by individuals;
- rake-off, where innovation begins to destabilize the existing ways
  of working;
- acceleration, where transformation occurs through socio-cultural,
  economic, ecological and institutional changes;
- stabilization, where the system reaches a new dynamic state of
  equilibrium.

Transformation is a growing field of thinking and practice for
adaptation. Tackling the impacts of more dangerous levels of climate
change will mean less incremental adjustment and more radical
decisions as whole areas become uninhabitable or agriculturally
unproductive. Mark Pelling (2010) presents a framework for making
sense of the range of choices facing humanity, structured around:
Resilience (which he interprets as attempts to maintain stability);
Transition (incremental social change and the exercise of existing
rights); and Transformation (the development of new rights claims and
changes in political regimes). His empirical analysis suggests that many
of the policy opportunities for transformational adaptive change have
not been seized upon to date (Pelling 2010). In their empirical study of
the Australian wine industry, Park *et al.* (2012) provide a useful
summary of some of the differences between incremental adaptation
actions and those that are more transformational (see Table 8.2).

Similarly, transformational changes are needed in the main sectors
driving greenhouse gas emissions. For example, currently it can take
decades for a new technology to become standard practice in the
power sector. This is far too slow to achieve the near-term reductions

Table 8.2 Incremental vs transformational adaptive change

Stage	Incremental	Transformational
Setting adaptation problem and context	Focus on reactive change management for the short-term, focusing on current conditions and finding ways to keep the present system in operation	Proactive management of present and future change that acknowledges future uncertainty and questions effectiveness of existing systems and processes
Adaptation processes	Limited source of info used to determine the need for, and the selection of, adaptation actions Climate can be managed by addressing sustainability more generally Lack of belief in anthropogenic climate change inhibits long-term change Perceived as having limited opportunities to implement change	Broad range of information used to determine the need for, and the selection of adaptation actions Climate change poses a distinct challenge and responses are differentiated across sectors and scales Directly managing for climate change will result in additional benefits
Evaluation and learning	Evaluation based on the current system and future is focused on surviving in the short-term	Continuous evaluation and monitoring of the system in response to all scales of change management Building capacity to create a totally new system or process, that is flexible and able to be acted upon if required

Source: Adapted from Park et al., 2012

in greenhouse gas emissions needed. Climate-smart technologies and management practices need to spread much more quickly. That requires moving away from prevailing linear approaches to supporting innovation – which see innovation as starting in a science lab and ending with diffusion to the market – to building effective 'innovation ecosystems' that recognize the complex and iterative, deeply uncertain, and heterogeneous nature of innovation processes and identify the critical services innovators need to thrive (Tawney et al., 2011). Those are the areas policy makers need to focus on when investing in innovation. Colleagues at the World Resources Institute have been working to develop a framework that provides step-by-step guidance to identifying the opportunities in the power sector and building a robust innovation ecosystem to capture and spread them globally (Tawney et al., 2011).

As noted in Box 8.5, ability to recognize and readiness to seize moments of opportunity as they arise will be key to driving transformation. An upshot of a world more prone to overlapping political, economic and ecological shocks is that it may provide key opportunities to press for transformational change. As Oxfam's Tim Gore wrote, from hurricane-battered New York in the immediate aftermath of Hurricane Sandy:

In the context in which an abrupt change of course is needed to address the
climate crisis – one some have compared only to mobilisation for war – crisis
moments can create unique windows of opportunity for non-linear political
change. That is precisely what we need. They can catalyse clear shifts in the
values and priorities of citizens, business and political leaders around the world.
Climate disasters in the global North and South alike are reminders of the
common threat we face, and of the need to act collectively and urgently to avert
yet greater harm.

(Gore, 2012)

Incremental and transformative approaches are not dichotomous
choices but complementary strategies. However, progressive forces in
society will need to invest much more in developing alternative,
transformational approaches and preparing shovel-ready ideas and
strategies in advance to take advantage of crises as opportunities for
their uptake.

Conclusion

Reconciling climate and development action in a world being
re-shaped by accelerating economic, demographic forces and global
ecological change requires nothing short of a fundamental re-ordering
of the relationship between humans and nature. Economic
development is needed to lift the living standards of the world's
poorest people but, paradoxically, achieving sustainable prosperity
requires finding alternatives to GDP-centered development strategies.
More public debate is also warranted about whether economic growth
is an appropriate goal for already rich countries.

To paraphrase John Maynard Keynes, the most difficult and
immediate challenge lies not so much in developing new ideas as in
wrenching ourselves from the clutches of old ones. While there are
profound uncertainties linked to the climate and global environmental
change more broadly, the biggest and most unsettling uncertainties we
face have to do not so much with science as with us humans. Will
communities and nations come together in time to address the climate
challenge with the degree of urgency and bold ambition that is
required?

Successful climate change management will require a dramatic scaling
up of mitigation and adaptation efforts at all levels, enabled by a
coordinated mix of policy and financial instruments. It will call for a

new development paradigm that mainstreams climate change into strategies and plans, and that links policy setting with the financing of solutions.

There is no silver bullet to solving the multiple interlocking crises of climate change, resources scarcities, inequality and poverty. But there are many changes that together add up to a solution. Although the search for the perfect approach or 'model' has great intellectual and political appeal, history tells us we are better off trying out a variety of ideas and approaches and learning from what works and can be replicated and scaled up. Diversification also allows for the spreading of risks.

Perhaps the most important lesson to draw from past development efforts is that development can never be a universal blueprint, and that successful development requires significant policy space, that is, the ability to experiment, learn and adapt policies to changing circumstances. This suggests that grand narratives of change be eschewed in favor of more eclectic and pragmatic responses to climate and development challenges suited to national/local circumstances.

Summary

- Development is at a crossroads – past successes do not provide a viable basis for addressing future challenges. Prevailing patterns of development present a seemingly inescapable trade-off between environmental sustainability and economic progress and have generated large social inequities.

- Sustained progress in advancing well-being in the twenty-first century will require development approaches that are more equitable, sustainable, and resilient. Alternative development paradigms include:
  - People-centered development
  - Green growth /green economy approaches
  - Steady-state economics
  - New environmentalist approaches

- While there is a wide range of differing views concerning the depth and scale of change needed, there is a good deal of convergence on basic conditions/requirements – or building blocks – for transitioning on to a sustainable development path:

    ○ A proactive government role

    ○ Transforming businesses

    ○ Better measuring the progress we want to see

    ○ Strengthened international cooperation

- Incremental and transformative approaches are not dichotomous choices but complementary strategies. To meet the challenges of development in a warming and resource-constrained world will ultimately require deep transformation of our economies and societies.

Discussion questions

- To what extent does the experience of advanced industrialized countries provide a model of prosperity that can be generalized, replicated or sustained?

- What are the essential features of climate-smart development?

- What measures can help ensure that a 'green economy' is also inclusive?

- What are the shortcomings of GDP as a measure of progress? How can it be improved?

- Is a strong public–private partnership the only option for delivering the kind of transformation needed to avert climate crisis? What should be the respective roles of government and businesses and how can they better collaborate?

Further reading

Brooks, N., Grist, N. and Brown, K. (2009) 'Development Futures in the Context of Climate Change: Challenging the Present and Learning from the Past', *Development Policy Review*, 27.6: 741–765.

304 • Alternative development futures

Daly, H. (2005) 'Economics in a Full World', in *Scientific American*, 293.3: 100–107.

Jackson, T. (2011) *Prosperity without growth: The transition to a sustainable economy*, Sustainable Development Commission.

McKinsey (2009) Pathways to a low-carbon economy. *Energy Policy*, 38.6: pp. 3067–3077. doi:10.1016/j.enpol.2010.01.047.

Moncel, R., Joffe, P., McCall, K. and Levin, K. (2011) *Building the Climate Change Regime: A Survey and Analysis of Approaches*, Working Paper, Washington D.C.: World Resources Institute.

O'Brien, K. (2012) 'Global environmental change II: From adaptation to deliberate transformation' *Progress in Human Geography* 36.5: 667–676.

OECD (2012b) *Inclusive Green Growth: For the Future We Want*, Paris: OECD.

Roberts, J.T. and Parks, B.C. (2007) *A Climate of Injustice: Global Inequality, North-South Politics and Climate Policy*, Cambridge, MA: MIT Press.

Stiglitz, J., Sen, A. and Fitoussi, J.-P. (2009) Report by the Commission on the Measurement of Economic Performance and Social Progress. SSRN Electronic Journal. doi:10.2139/ssrn.1714428.

UNEP (2011b) *Towards a Green Economy – Pathways to Sustainable Development and Poverty Eradication*, UNEP.

Wilkinson, R. G. and Pickett, K. (2010) *The spirit level: Why greater equality makes societies stronger*, New York: Bloomsbury Press.

Websites

www.unep.org/greeneconomy/
The United Nations Environment Programme (**UNEP**) **Green Economy Initiative** website is a rich resource for research, news, webinars and success stories from around the world on building a green economy. The aim is to provide the analysis and policy support for investing in green sectors and in greening environmental unfriendly sectors.

www.wavespartnership.org/waves/
The Wealth Accounting and the Valuation of Ecosystem Services (**WAVES**) is a global partnership that aims to promote sustainable development by ensuring that the national accounts used to measure and plan for economic growth include the value of natural resources. The WAVES partnership includes UN agencies, governments, international institutes, NGOs and academics to implement environmental accounting where there are internationally agreed standards, and develop standard approaches for other ecosystem service accounts.

http://steadystate.org/category/herman-daly/
The website of the **Center for the Advancement of the Steady State Economy** (**CASSE**) is a good place to go for a range of educational resources and policy briefs addressing key concepts as well as means to promote the steady-state economy as a desirable alternative to GDP growth.

 Conclusion

This generation's challenge: Piloting spaceship Earth.

⦿ **State of play**
   ◌ **The science**
   ◌ **The action**
⦿ **Five challenges for this decisive decade – charting the critical path**
   1 Bending the GHG emissions curve
   2 Getting ready for a 4°C world
   3 Igniting the resource productivity revolution
   4 Making cities sustainable, liveable and resilient
   5 A new global partnership for sustainable development

> We are now faced with the fact, my friends, that tomorrow is today. We are confronted with the fierce urgency of now. In this unfolding conundrum of life and history, there is such a thing as being too late. Procrastination is still the thief of time. Life often leaves us standing bare, naked, and dejected with a lost opportunity. The tide in the affairs of men does not remain at flood – it ebbs. We may cry out desperately for time to pause in her passage, but time is adamant to every plea and rushes on. Over the bleached bones and jumbled residues of numerous civilizations are written the pathetic words, 'Too late.'
>
> Martin L. King, Jr.

The decade ahead will be decisive for tackling **climate change**: failure to bend the curve downwards on **GHG** emissions by 2020 will likely lead to runaway climate change with potentially disastrous consequences for humanity. As we have seen in Chapter 4, delaying action will only make the inevitable adjustments more difficult and costly, in the same way that swimming against the current becomes more difficult the closer one gets sucked towards the edge of a waterfall.

At the same time, ours is the first generation in history to hold within its grasp the prospect of a world rid of the scourge of extreme poverty and in which everyone can go to bed on a full stomach.

A central message of this book is that these two imperatives – climate **stabilization** and the eradication of poverty and hunger – are intimately intertwined and interdependent. They can only be delivered in a coherent and mutually reinforcing manner. Climate change is a fundamental threat to economic development and poverty reduction. Failure to address climate change will put in jeopardy future development efforts and bequeath a depleted and degraded Earth to our children and future generations. Conversely, it is only by pointing the way to climate stabilization pathways that are compatible with development progress that a global deal will be reached to avert the **climate** crisis. Any deal that doesn't respect and support poor countries' right to develop is dead on arrival.

This book as a whole was written with the future in mind: as authors we have strived to look beyond the immediate concerns of the present day, to present main generic issues and analytical frameworks that can be applied to understanding the climate–development nexus as it evolves over time.

This concluding chapter, however, returns to the near present to take stock of where we are in terms of the science and actions taken, and to consider the critical challenges that must be tackled in this decisive decade.

1. The current state of play

The scientific debate about anthropogenic climate change has been settled. As noted in Chapter 1, the overwhelming weight of scientific consensus has established that climate change is happening and that global warming is caused by human factors, namely the burning of fossil fuels, deforestation and land-use changes. Certainly, uncertainties remain within climate science: we have much to learn for example about the role and behavior of clouds, oceans and ice sheets, and potential feedback effects that may accelerate or slow global warming. However, uncertainty cuts both ways. Things *could* turn out better than current scientific consensus cautions, but they could also be much worse. The grounds for optimism are thin: where new data and inquiry show previous scientific findings to be deficient, it is usually in the sense of underestimating the gravity or speed of climate change. As climate science advances we tend to discover more positive feedback loops that accelerate global warming. We have yet to find confirmation of a significant negative feedback that could slow global warming.

There remain important gaps in knowledge about the complex interactions between human and earth systems. Chapter 7 describes the ever-closer entanglement of socio-economic and environmental processes. It highlights the global scale, speed and growing complexity of change. Yet the knowledge that informs responses to the interconnected climate and resource challenges we face still tends to be from sub-disciplinary silos and rooted in reductionist and mechanistic paradigms of change. The problem cannot just be pinned on the scientific/academic community. Funding and demand for science encourages the production of specialized rather than holistic knowledge. There is as a result a big gap in evidence and knowledge to help pilot mother-ship Earth within the planet's 'safe operating zone'. With improvements in our ability to manage and process large data sets ('big data'), however, this gap can be quickly closed.

Globally, we are falling far short in responding to the climate challenge. To date, our commitments have not measured up to what the scientific establishment tells us is needed to avoid 'dangerous' climate change, and yet our actions have fallen short of even those commitments. **Greenhouse gases** in 2012 were 58 per cent above 1990 levels and 14 per cent above where they need to be in 2020 for us to have a chance of keeping temperature increases below 2°C (UNEP, 2012). The international momentum that had gathered ahead of the 2009 UN climate conference in Copenhagen has largely fizzled out with the global economic crisis. Atmospheric concentrations and emissions of greenhouse gases continue to increase, with CO_2 concentrations now reaching 400ppm and emissions at an all-time high of 35.6 billion tonnes in 2012 (Peters *et al.*, 2012). The last time atmospheric concentrations were so high was three million years ago in the Pliocene epoch.

Alarmingly, the trend over the past decade has been acceleration in emissions: according to figures published in the journal *Nature Climate Change*, the average annual increases in global CO_2 levels were 1.9 per cent in the 1980s, 1.0 per cent in the 1990s and 3.1 per cent since 2000 (Peters *et al.*, 2012). It is likely that we have now missed the window for keeping global warming below 2°C. Instead, we are on track to warm by 3–4°C by the end of the century and extreme weather will become the 'new normal', affecting every region in the world (Anderson and Bows, 2008; World Bank, 2012d).

Against this gloomy backdrop there are some bright spots of hope. Even without a binding global **UNFCCC** agreement, there are numerous examples of leadership and bold action by countries, companies, cities and civil society, often networked trans-nationally. China has targets to reduce the carbon intensity of its economy, is investing more than any other country in renewable energies and has started experimenting with carbon emissions trading domestically. South Korea introduced a 'green stimulus' package to re-launch its economy onto a greener path. Germany increased the share of renewable sources in its energy mix from 6.4 per cent in 2000 to 20 per cent in 2011.

Cities are also starting to organize and lead the way in the fight against climate change. On the eve of the Rio+20 Conference, a group of mayors around the globe announced steps to slash emissions of greenhouse gases by 248 million metric tonnes in 2020, and by more than 1 billion tonnes by 2030. The C40 Leadership Group, which includes 59 cities producing about 14 per cent of the world's greenhouse gas emissions, have undertaken nearly 5,000 climate-related actions since the network first formed in 2005. New York, for example, has plans to reduce municipal greenhouse gas emissions by 30 per cent from 2005 levels by 2030, while Bogota plans to cut by 16 per cent from 2007 levels by 2019 (www.c40cities.org).

Climate change considerations are increasingly mainstreamed into development planning and aid. A growing number of developing countries are integrating low carbon and climate resilient considerations into their national development plans (e.g. Rwanda, Ethiopia), while most leading international development agencies have created strategies to tackle climate change and address climate risks across their development cooperation portfolios (Hammill and Tanner, 2011).

Businesses are also contributing to the fight against climate change. Global investment in renewable energy is already approaching global investments in fossil fuels despite the lack of a global carbon price and the continuation of massive fossil fuel subsidies. The global market for clean technology manufacturing doubled between 2008 and 2011 to reach almost €200 billion per year. Big multi-national companies are also increasingly factoring climate impacts, resource risks and **adaptation** into their operations, strategies and supply chain management.

At the international level, in the vacuum of leadership left by national governments, plurilateral 'clubs' and transnational networks of action have emerged and are rapidly growing to be important drivers of action and ambition (Bulkeley and Newell, 2010; Moncel et al., 2011). The C40 Cities Climate Leadership Group, the Green Growth Action Alliance (G2A2) and the Sustainable Energy for

All initiative are cases in point. The challenge going forward will be to combine that groundswell of action into a comprehensive, inter-governmentally negotiated **regime**, or at least to create the space within the regime for these actions to be recognized and supported.

2. Five challenges for this critical decade

2.1 Bending the GHG emissions curve

As outlined in Chapter 3, the international climate talks in Doha in 2012 created a bridge between the old climate regime – the Kyoto regime, that will expire in 2020 – and a future one whose shape, ambition and contents are yet to be determined. Importantly, the talks established a single negotiations platform with a commitment to agreeing a new comprehensive agreement by 2015. Sealing an equitable and ambitious climate deal in 2015 will be a crucial step towards galvanizing international action to bend the greenhouse gas emissions curve downwards.

This creates an important opportunity to rethink and reframe the climate deal in light of current realities, to re-infuse it with a sense of urgency and moral purpose. The 2015 climate negotiations in Paris will provide an opportunity to build an international climate regime that reconciles concerns of **equity** with enhanced ambition to close the emissions gap and catalyze low-carbon, climate resilient development. Achieving this objective will require a climate deal in 2015 that has the following features:

(a) **Ambition**: it must include ambitious **mitigation** pledges in line with a clear overall stabilization target and increased ambition across the different pillars of the regime (adaptation, finance, technology, capacity building, and architecture).

(b) **Equity**: to achieve the level of ambition needed to avert dangerous climate change will require rethinking and operationalizing the principle of equity within the international climate regime, such that it drives rather than inhibits collective action. In the place of retrospective and potentially negotiation-blocking discussions on equity as burden sharing, a narrative is needed that: (a) roots equity in mutuality, reciprocity, solidarity, and the protection of the most vulnerable people; (b) embeds equity concerns across the pillars of the new regime; and (c) frames ambition as a means for achieving equity.

(c) **Trust**: a series of confidence-building measures will be needed to restore trust between negotiating parties to a level where meaningful collective actions can be agreed. Strengthening the transparency and accountability of the UNFCCC framework and associated instruments will be crucial in that regard.

(d) **Effective lobbying**: Ensuring that the swath of activities outside national government policy or UNFCCC commitments are brought to bear on the negotiations as a signal of enthusiasm and support for a global agreement.

The centrality of the equity issue is eloquently explained by Edward Cameron (2013) at the World Resources Institute:

> In the new climate agreement, equity cannot be about sharing failure. It must become a means to share both the opportunities and challenges of the transition to climate-compatible development. In addition, equity cannot remain a quarrel about the past. It must be our opportunity to secure a fair future for all, with equitable access to sustainable development and respect for planetary boundaries.

While redoubling efforts to build an equitable and ambitious new climate regime, parallel and complementary actions will also be required outside of the UNFCCC framework particularly in the period between 2015 and 2020 when the new climate deal comes in to effect. Plurilateral approaches will be important for getting major emitters around the table to agree ambitious mitigation targets. Ambitious partnerships should be formed to aggressively pursue big win-win opportunities between climate and development, such as restoring degraded lands and expanding access to low-cost low-carbon energy. Equally critical will be to take steps towards putting a predictable global price on carbon so that the full power of the market and private entrepreneurship can act to deliver emissions reductions at scale. The first major step in that direction would be to phase out fossil fuel subsidies. That alone could lead to a 5 per cent reduction in emissions by 2020 according to the Natural Resources Defense Council (NRDC, 2012).

2.2 Getting ready for a 4°C world

Given the failure of action to stem the rise in emissions and the lack of an agreed target for stabilization of atmospheric greenhouse gas concentrations, many scientists now suggest that the adaptation challenge needs an upgrade. While the 2°C stabilization target remains an important aspirational target for mitigation, policy makers need to factor in the very real possibility that future climate change will be much more severe than implied by the 2°C threshold. On current efforts there is roughly a 20 per cent likelihood of exceeding a 4°C global temperature rise by 2100, while this could happen by the 2060s if pledges are not honored (World Bank, 2012d). This requires planning for unprecedented heat waves, severe drought, major floods, sea level rise and ecosystem changes such as forest and coral die-back. As such, learning about and investing in adaptation is required at a far greater scale, based less on incremental adjustment and more on a reassessment of whether human and ecological systems can continue to deliver services and welfare.

2.3 Igniting the resource revolution

As the global economy doubles and the global middle class grows by 3 billion by 2030, nothing short of a resource revolution will be needed to keep pace with the soaring demand for natural resources and to avert an era of higher and more volatile resource prices.

The resource revolution rests on two other 'R's:

(a) *Reducing* the amount of natural resources needed to produce any given level of economic output. This requires radically increasing resource use efficiency (also termed 'resource productivity'); minimizing our dependence on non-renewable sources of energy and materials by scaling up renewable alternatives; and incentivizing the reuse of waste streams as productive inputs through industrial ecology (also known as 'industrial symbiosis') approaches.
(b) *Restoring* and sustaining ecosystems to rebuild the natural capital stock. This involves environmental investments in rehabilitating degraded natural assets such as forests, grasslands, wetlands, marshes, mangroves, lakes, rivers and coral reefs. Restoring degraded ecosystems can generate substantial and multiple economic and social returns as well.

Radical increases in resource productivity are within reach. Assessing opportunities across all four key resources – energy, land, materials, and water – McKinsey found that available productivity measures could address up to 30 per cent of total demand in 2030 (Dobbs, 2011). The total value to society associated with these opportunities is up to US\$3.7 trillion, with 70 per cent of these opportunities having returns above 10 per cent, even based on today's prices and without considering any carbon taxes (*ibid*). As we have seen, big multinational companies – such as Unilever, SABMiller, Nestlé, General Electric and Siemens – are already waking up to these opportunities by investing heavily in energy and water efficiency across their supply chains. If famed venture capitalist Vinod Khosla is right, disruptive technological innovation will 'completely upend assumptions in oil, electricity, materials, storage, agriculture, and the like' (Khosla, 2011).

The opportunities and potential benefits of restoring degraded ecosystems are every bit as compelling, from a hardnosed business perspective as well as from social and environmental perspectives. The world's ecosystems generate services to humanity worth over US$70 trillion per year, which is more than the entire world GDP as of 2013 (Sukhdev *et al.*, 2010). Given that about two-thirds of the world's ecosystems are already degraded, restoration must be given high priority. Far from acting as a brake on growth and development, restoring lost and damaged ecosystems can trigger multi-million dollar returns, generate jobs and play a vital role in reducing poverty around the world (Nellemann and Corcoran, 2010).

A key to igniting the resource revolution will be to strengthen price signals for natural resources so that market forces and private entrepreneurialism can be unleashed and put to work increasing resource productivity and restoring ecosystems. In the words of Gus Speth: 'our market economy is operating on wildly wrong market signals, lacks other correcting mechanisms, and is thus out of control environmentally' (Speth, 2008). Private actors will only have the confidence to invest in low carbon, resource saving and clean technologies if natural capital is correctly valued in market prices and transactions.

Market incentives remain tilted in favor of fossil fuel energy and wasteful resource use: subsidies to energy, water and agriculture add up to US$1 trillion per year, more than 12 times the amount of support for renewable energy. Correcting market prices involves two important steps: first, eliminating distorting fossil fuels and resource use subsidies; second, adjusting prices to incorporate external costs. This will involve tackling head-on the powerful economic and political interests vested in the fossil-energy system and extractives sector. Doing so will free up massive fiscal and financial resources that can be used to compensate those who are most severely impacted by price increases, including through targeted cash transfers to ensure poor people are not unduly penalized. Part of the proceeds from carbon and resource taxes can also be channeled back into accelerating the resource revolution, for example by supporting clean and renewable energy research and innovation.

2.4 Making cities sustainable, liveable and resilient

Cities will be one of the decisive arenas in the drive to protect the climate and eradicate poverty. Given the rapid rate and massive scale of urbanization (discussed in Chapters 4 and 7), successfully tackling climate change and development challenges in cities is crucial. Cities generate 80 per cent of world GDP and account for a similar share of global greenhouse gas emissions. They are engines of growth, hubs of innovation and provide the best available means to absorb growing pools of labor as global population grows. By concentrating people, resources, wealth and pollution sources, rapid urbanization provides unprecedented opportunities to achieve economies of scale in deploying low carbon technologies, and to build sustainable cities through investment in efficient infrastructure, inclusive public services, and urban forms that have the lowest possible resource intensity.

The stakes could not be higher. Cities are the stage on which the most intense social and environmental struggles are being played out. Rapid, unplanned and uncoordinated urban development can result in the explosion of slums, crumbling infrastructure, entrenched inequities, spiraling congestion and pollution problems.

Vulnerability to climate impacts is likely to increase over the next two decades as the land area occupied by cities triples to 600,000 square kilometers, often in disaster-prone sites and without basic infrastructure and policies to mitigate disaster risks (Mohieldin and Allaoua, 2013).

Since cities are difficult to reshape once they are built, it is imperative that climatic risks and environmental and social concerns are integrated into urban development plans at the earliest possible stage. Repeating the mistakes of many sprawling cities in rich countries that are more suited for cars than for people will create **lock-in** to a future of vulnerability, rising oil dependence, severe urban air pollution and rising greenhouse gas emissions for possibly hundreds of years to come.

There is no silver bullet, but building sustainable, liveable and resilient cities will entail at a minimum the following (World Bank, 2011):

● Effective land management and urban planning – this requires land and housing regulations and market-based incentives that promote urban densification and encourage compact cities.
● Government mandates and policies (at national or municipal level) to increase energy efficiency and promote cleaner energy sources for electricity generation, buildings, and urban transport.
● Development of public and multi-modal transportation systems that allow low-cost and low carbon mobility and offer viable alternatives to automobile use.
● Contingency planning for potential disruptions in water supply associated with climate change.
● Investments in efficient infrastructure and basic service provision.

Plate 9.1 *Two sides of urbanization: (a) Slums of Rio; (b) Construction in Dubai, UAE.*

(a)

(b)

Credit: (a) WRI /Crystal Davis; (b) WRI /James Anderson.

Note: Rapid urbanization creates historic challenges as well as opportunities for development and climate action. Effective land management and urban planning and the development of public and multi-modal transportation options will be crucial to avoid the expansion of slums and urban poverty as massive migration to urban areas occurs in the next decades. Rapid development of urban infrastructure will also create huge economies of scale for deploying clean and energy efficient technology. Ensuring new buildings are built with energy efficiency in mind is one of the most effective measures for reducing energy consumption and related greenhouse gas emissions in cities.

As an increasing percentage of the world's population moves to urban areas, the authority and capacity of city institutions to shape the relationship between environment and development will also continue to grow. Good local governance and institutions, management and planning will underpin the success of cities in meeting the climate change and development challenge.

2.5 A new global partnership for sustainable development

As the target period of the Millennium Development Goals reaches its end in 2015, frenzied multi-level deliberations are underway to define the next set of international development goals. The shaping of the post-2015 development goals framework offers an important opportunity for putting the issues of sustainability, equity and development in the context of a broader and more fundamental

renegotiation of the international 'social contract'. In the words of
O'Brien and colleagues (O'Brien *et al.*, 2009: 1):

> Global social–ecological change is creating new challenges and opportunities for
> both states and citizens, inevitably forcing a rethinking of existing and evolving
> social contracts in the light of ecosystem changes, more extreme weather events,
> and the consequences of social–ecological changes in distant locations.

Proposals for the development of a set of international Sustainable
Development Goals (SDGs) were at the heart of discussions at the
Rio+20 Summit in 2012. The SDGs present an opportunity to reframe
the next generation of international development goals in the context
of planetary stewardship, putting global sustainability at its very core.
This meeting agreed to set up a process to develop a set of SDGs that
would be 'coherent with and integrated into the United Nations
development agenda beyond 2015 . . . [but] not divert focus or effort
from the achievement of the Millennium Development Goals'
(UNGA, 2012: 47).

If it is to lead to a breakthrough for sustainability, the post-2015 global
partnership for sustainable development must be driven by a shared
vision and a strong common sense of moral purpose. Four sets of
considerations will be particularly important in determining whether
the new post-2015 development agenda meets the challenges of our
times:

1   **Is it forward-looking and 'future-fit'?** While there will be some
    important unfinished business from the MDGs to deal with, it is
    important that the new set of goals be designed to tackle
    tomorrow's problems, not yesterday's. The world in 2030 will
    look very different from today, let alone when the MDGs were
    conceived. Poverty will be more urban and concentrated in fragile
    states. Power will be more diffuse in the international system.
    Human, economic and ecological systems will be more tightly
    enmeshed. Issues of access rights and equity will gain prominence
    as scarcities start to bite. Climate risks and environmental
    pressures will intensify, spurring a stronger imperative to
    coordinate global action to address climate change and avert
    ecological disaster.
2   **Are equity and sustainability embedded in the new goals'
    architecture, the means of implementation and the metrics of
    progress?** Integrating social, environmental and economic

dimensions of sustainable development will require explicit and specific goals centered on environmental sustainability and social inclusion. Beyond that it will also require equity and sustainability concerns to be woven into other relevant goals and their implementation. Finally, new metrics will be needed to measure progress, enable learning and foster transparency and accountability.

3   **Is it universal in scope?** The challenges of global development today call for global solutions and coordinated international action. In our interconnected world marked by growing economic and ecological interdependence, the security and development prospects of every country, rich and poor alike, are increasingly tied to the effective governance of **global public goods**. The moment is here for a single, shared development agenda that applies to all countries while being sensitive to different national realities, capacities and levels of development.

4   **Will it be implemented through a global partnership?** For a universal agenda to have traction it will need to be backed by a broad-based global partnership, that entails commitments, responsibilities and accountability from states – rich and poor – as well as from important non-state actors such as businesses, financial investors, and private foundations. This will be in marked difference to the MDGs, which mainly involved a compact between governments entailing commitments from rich countries to increase development aid and commitments from developing countries to use this aid to advance the MDGs.

The development landscape is being transformed by climate change in ways that call for a fundamental rethink of development practices, policies and pathways. While the challenges ahead are formidable, we have never been better equipped to face them. There are significant opportunities in meeting the linked challenges of reducing emissions, preparing for the impacts of a changing climate, building sustainable cities, and forging a new global partnership for sustainable development. This book is a response to this urgent call: we hope that with it readers will be better equipped with analytical tools and improved understanding to help steer spaceship Earth to meet the mother of all development challenges.

> We travel together, passengers on a little space ship, dependent on its vulnerable reserves of air and soil; all committed for our safety to its security and peace; preserved from annihilation only by the care, the work, and, I will say, the love

we give our fragile craft. We cannot maintain it half fortunate, half miserable, half confident, half despairing, half slave—to the ancient enemies of man—half free in a liberation of resources undreamed of until this day. No craft, no crew can travel safely with such vast contradictions. On their resolution depends the survival of us all.

Adlai Stevenson, US ambassador to the United Nations, 1965

Summary

- The scientific argument about human-made climate change is all but over, but blind spots remain in our knowledge.

- There has been much activity to mitigate the causes of and adapt to the impacts of climate change, despite the limitations of the international climate regime.

- Five challenges for this decisive decade remain:

  1   Bending the GHG emissions curve

  2   Getting ready for a 4°C world

  3   Igniting the resource productivity revolution

  4   Making cities sustainable, liveable and resilient

  5   Forging a global partnership for sustainable development post-2015.

 **Glossary**

Adaptation
Adjustment in natural or human systems in response to actual or expected climatic stimuli or their effects, which moderates harm or exploits beneficial opportunities (IPCC, 2001a).

Adaptation deficit
The adaptation deficit is a term used to describe the inability to cope with shocks and stresses imposed by the existing variability of the climate, even before additional climate change impacts are taken into account. Taking actions to close the pre-existing adaptation deficit is regarded by many as the priority step for many developing countries where it involves '**no regrets**' **or** '**low regrets**' **measures** that will be beneficial irrespective of future climate change.

Adaptive capacity
The combination of the strengths, attributes and resources available to an individual, community, society, or organization that can be used to prepare for and undertake [**adaptation**] actions to reduce adverse impacts, moderate harm, or exploit beneficial opportunities. (Source: IPCC, 2012)

Additionality
Additionality is referred to in the context of climate finance and climate change mitigation policy, particularly the Clean Development Mechanism (CDM).

In climate finance, additionality relates to the pledge for 'new and additional resources' referred to in UNFCCC texts, particularly since 2009. Its definition is politically contentious, with many developing countries seeking an increase in climate finance that is not connected to existing aid flows (ODA), or at least climate finance that is over and above the existing pledged targets for aid of 0.7% of rich countries' gross national income. Some richer countries have argued that 'additional' climate finance should include any increases in ODA spent on climate change activities since pledges were made at the Copenhagen meeting in 2009 (Brown *et al.*, 2010).

For the CDM, additionality refers to ensuring that emission reductions are only to be certified under the CDM if they are additional to any that would occur in the absence of the project activity. Additionality is currently demonstrated if project proponents can show that realistic alternative scenarios to the proposed project would be more economically attractive with CDM finance or that the project faces barriers that CDM helps it overcome.

Carbon budget/carbon space

The term 'carbon budget' is used to refer to the total amount of carbon dioxide and other greenhouse gases emissions available over time in order to meet a given stabilization target. It is often used interchangeably with 'carbon space', although strictly speaking, the latter term refers to the portion of the carbon budget that is as yet unused. Carbon space is often used in discussions regarding the division of this total amount in accordance with development needs of all people on the planet, and particularly those advocating a per capita emissions allocation approach as an equitable solution.

It should be noted that in climate science the term 'global carbon budget' is also used in a different sense, to refer to the balance of the exchanges (incomes and losses) of carbon between the carbon reservoirs in the carbon cycle.

Climate

Climate is usually defined as the 'average weather', or more rigorously, as the statistical description of the weather in terms of the mean and variability of relevant quantities over periods of several decades (typically three decades as defined by the World Meteorological Organisation). These quantities are most often surface variables such as temperature, precipitation, and wind, but in a wider sense the 'climate' is the description of the state of the climate system. (Source: IPCC, 1995b)

Climate change

Climate change refers to a change in the state of the **climate** that can be identified (e.g., by using statistical tests) by changes in the mean and/or the variability of its properties, and that persists for an extended period, typically decades or longer. Climate change may be due to natural internal processes or external **forcings**, or to persistent anthropogenic changes in the composition of the atmosphere or in land use. Note that the **United Nations Framework Convention on Climate Change (UNFCCC)**, in its Article 1, defines climate change as: 'a change of climate which is attributed directly or indirectly to human activity that alters the composition of the global atmosphere and which is in addition to natural climate variability observed over comparable time periods'. In contrast to the IPCC definition, the UNFCCC thus makes a distinction between climate change attributable to human activities altering the atmospheric composition, and climate variability attributable to natural causes. (Source: IPCC, 2007)

Disaster Risk Reduction (DRR)

Denotes both a policy goal or objective, and the strategic and instrumental measures employed for anticipating future disaster risk; reducing existing exposure, hazard, or vulnerability; and improving **resilience** (IPCC, 2012). DRR builds on disaster risk management (DRM), which describes processes for designing, implementing, and evaluating strategies, policies and measures to improve the understanding of disaster risk, foster disaster risk reduction and transfer, and promote continuous improvement in disaster preparedness, response, and recovery practices, with the explicit purpose of increasing human security, well-being, quality of life, and sustainable development (IPCC, 2012).

Equity and social justice

Equity is an ethical principle based on the notion of moral equality between individuals, i.e. that all should be treated as equals. Equity has three main dimensions: (a) equality of opportunity – a person's life chances should not be determined by the accident of birth; (b) equal concern for people's needs – ensuring that all are guaranteed basic rights to access goods and services fundamental to their needs and to preserving their human dignity; (c) meritocracy – that is,

positions in society and rewards reflect differences in effort and
ability and are based on fair competition (Jones, 2009).

Social justice is best understood as a process, rather than an end
point or outcome. More specifically, it is a process that (a) seeks a
fair distribution of resources, opportunities and responsibilities;
(b) challenges the roots of oppression and injustice; (c) empowers
all people to fulfil their potential and lead lives they value;
(d) builds social solidarity and capacity for collective action
(adapted from UC Berkeley 7th Annual Justice Symposium).
Two key aspects therefore are social inclusion and individual
empowerment.

Forcings

The climate system evolves in time under the influence of its own
internal dynamics and due to changes in external factors that affect
climate (called 'forcings'). External forcings include natural
phenomena such as volcanic eruptions and solar variations, as well as
human-induced changes in atmospheric composition. Solar radiation
powers the climate system. There are three fundamental ways to
change the radiation balance of the Earth: (1) by changing the
incoming solar radiation (e.g. by changes in Earth's orbit or in the sun
itself; (2) by changing the fraction of solar radiation that is reflected
(called 'albedo': e.g. by changes in cloud cover, atmospheric particles
or vegetation); and (3) by altering the longwave radiation from Earth
back towards space (e.g. by changing greenhouse gas concentrations).
(Source: IPCC, 2007)

Global public goods

In economics, the term 'public good' refers to a good that produces
benefits that cannot be easily restricted to one consumer or set of
consumers. A public good possesses two defining characteristics:
(1) it is *non-rival*, meaning that its consumption or enjoyment by one
person does not negatively affect others' consumption or enjoyment of
the same good (think of knowledge, or broadcast television), (2) it is
non-excludable, meaning that it is impossible (or very difficult and
costly) to exclude anyone from consuming it (think of street lights or a
lighthouse).

Such goods are usually under-provided or not provided at all by
private agents in the market, not least since once provided they can be
enjoyed for free (the 'free rider' problem). As a result, public
intervention (in the form of regulation, public financing or direct

provisioning) is usually required to ensure public goods are provided at a socially optimal level.

Although there is no agreed definition of 'global public goods', this is generally applied to public goods that provide benefits that are global in scope, such as international peace, nuclear non-proliferation and climate stability. Providing such goods typically requires some form of international cooperation.

Greenhouse effect

Greenhouse gases effectively absorb thermal infrared radiation, emitted by the Earth's surface, by the atmosphere itself due to the same gases, and by clouds. Atmospheric radiation is emitted to all sides, including downward to the Earth's surface. Thus greenhouse gases trap heat within the surface-troposphere system. This is called the **greenhouse effect**. Thermal infrared radiation in the troposphere is strongly coupled to the temperature of the atmosphere at the altitude at which it is emitted. In the troposphere, the temperature generally decreases with height. Effectively, infrared radiation emitted to space originates from an altitude with a temperature of, on average, $-19°C$, in balance with the net incoming solar radiation, whereas the Earth's surface is kept at a much higher temperature of, on average, $+14°C$. An increase in the concentration of greenhouse gases leads to an increased infrared opacity of the atmosphere, and therefore to an effective radiation into space from a higher altitude at a lower temperature. This causes a radiative forcing that leads to an enhancement of the greenhouse effect, the so-called enhanced greenhouse effect. (Source: IPCC, 2007)

Greenhouse gas (GHG)

Greenhouse gases are those gaseous constituents of the atmosphere, both natural and anthropogenic, that absorb and emit radiation at specific wavelengths within the spectrum of thermal infrared radiation emitted by the Earth's surface, the atmosphere itself, and by clouds. This property causes the **greenhouse effect**. Water vapour (H_2O), carbon dioxide (CO_2), nitrous oxide (N_2O), methane (CH_4) and ozone (O_3) are the primary greenhouse gases in the Earth's atmosphere. Moreover, there are a number of entirely human-made greenhouse gases in the atmosphere, such as the halocarbons and other chlorine- and bromine-containing substances, dealt with under the Montreal Protocol. Beside CO_2, N_2O and CH_4, the Kyoto Protocol deals with the greenhouse gases sulphur hexafluoride (SF_6), hydrofluorocarbons (HFCs) and perfluorocarbons (PFCs). (Source: IPCC, 2007)

Hazard

The potential occurrence of a natural or human-induced physical event that may cause loss of life, injury, or other health impacts, as well as damage and loss to property, infrastructure, livelihoods, service provision, and environmental resources (IPCC, 2012).

Environmental and climate hazards occur at the intersection of natural/climatic and human systems. It is useful to distinguish between climate hazards, climate disasters and extreme weather events. While climate hazards are the *possibility* of a dangerous and harmful event occurring, climate disasters refer to the actual occurrence of an event that causes damage when hazards combine with exposed systems (for example, people living on low-lying areas prone to flooding). Extreme weather refers to weather events that are outside of normally observed weather variations, and do not necessarily always cause harm to humans.

Human development

Human development is a process of enlarging people's choices. In principle, these choices can be infinite and change over time. But at all levels of development, the three essential ones are for people to lead a long and healthy life, to acquire knowledge, and to have access to resources needed for a decent standard of living. If these essential choices are not available, many other opportunities remain inaccessible. But human development does not end there. Additional choices, highly valued by many people, range from political, economic and social freedom, to opportunities for being creative and productive, and enjoying personal self-respect and guaranteed human rights.

Human development has two sides: the formation of human capabilities (such as improved health, knowledge and skills) and the use people make of their acquired capabilities (for leisure, productive purposes or being active in cultural, social and political affairs). If the scales of human development do not finely balance the two sides, considerable human frustration may result. According to this concept of human development, income is clearly only one option that people would like to have, albeit an important one. But it is not the sum total of their lives. Development must, therefore, be more than just the expansion of income and wealth. Its focus must be people. (Source: UNDP, 1990)

Lock-in effect

The longevity and inertia of many infrastructural investments – such as in power plants, building stocks and urban expansion – can prolong the operation of obsolete technologies that are dependent on carbon-rich fuels and/or inefficient in energy use, thus creating large-scale and long-term commitment to carbon intensive pathways. This is known as the carbon lock-in effect.

Carbon lock-in can also result from the existence of barriers to developing and deploying clean alternative energy technologies on a large scale. For example, the absence of a global carbon price may limit incentives to invest in climate-friendly technologies.

The concept of carbon lock-in is most used in relation to the challenges of expanding energy supply and urban infrastructure in a manner consistent with climate protection.

'Low-carbon' or 'low-emissions' development

Development that results in lower emissions of GHGs compared with the baseline, business-as-usual development path.

Mainstreaming

Climate mainstreaming is the integration of the climate change considerations into every stage of a development planning, policy or project process (from design, to implementation, monitoring and evaluation), with the aim of minimizing climate-related risks and promoting low-emissions development options.

Maladaptation

Maladaptation can be defined as action taken ostensibly to avoid or reduce vulnerability to climate change that impacts adversely on, or increases the vulnerability of, other systems, sectors or social groups. (Source: Barnett and O'Neill, 2010)

Market failure

The term market failure has a very specific meaning in economics: it refers to a situation where the free interplay of private agents fails to result in resources being allocated efficiently (i.e. in a socially optimal way) due to a breakdown of the price mechanism. This in turn is caused by the existence of externalities, public goods, informational asymmetries and the lack of clearly defined property rights over common pool resources (Sterner, 2003). Market failures result in a misallocation of resources because private incentives are at odds with what is optimal from a societal perspective. Nick Stern and his review team characterized climate change as the greatest and most wide-ranging market failure the world has ever seen (Stern, 2007).

Mitigation

Climate change mitigation refers to actions taken to reduce the rate at which **greenhouse gases** accumulate in the atmosphere, in order to minimize climate change and its effects. This is done by reducing greenhouse gas emissions or removing greenhouse gases from the atmosphere through enhancing carbon sinks (e.g. trees, soil, vegetation).

Nationally Appropriate Mitigation Actions (NAMAs)

As part of the agreed outcomes of the Bali Action Plan concluded at COP 18 in Doha, developing countries will take Nationally Appropriate Mitigation Actions (NAMAs) in the context of sustainable development, supported and enabled by technology, financing and capacity-building, aimed at achieving a deviation in emissions relative to 'business as usual' emissions in 2020. NAMAs refer to sets of policies and actions that countries undertake as part of a commitment to reduce greenhouse gas emissions. These NAMAs are diverse and range from project-based mitigation actions to economy-wide emission reduction objectives. The term recognizes that different countries may take different nationally appropriate action on the basis of equity and in accordance with common but differentiated responsibilities and respective capabilities. Reporting and communication of NAMAs by developing countries is also designed to facilitate the matching of finance, technology and capacity-building support with these actions.

Negative emissions

Negative emissions include burning biomass and capturing the carbon released to store it underground (a technology not yet proven on a commercial scale), large-scale afforestation (which has potentially very significant impacts on foods supply), increasing the carbon content of soil through bio-char, and capturing carbon from the air and storing it underground.

Producing negative emissions can have undesirable side effects. In particular, adverse impacts on biodiversity have raised concern. The recent UN Convention on Biological Diversity agreed on a de facto moratorium on large-scale geo-engineering activities that could negatively impact on biodiversity, which includes large-scale biomass with carbon capture and storage, large-scale reforestation and large-scale production of bio-char for adding to soil. (Source: Friends of the Earth, 2011)

No-regret/low-regret measures
Measures to deal with the **adaptation deficit** are referred to as **no regret** options because they are not affected by uncertainties related to future climate change. For example, improved provision and dissemination of early warning systems will be beneficial irrespective of how the climate changes. Some related adaptation measures are referred to as **low-regret** options as they require some investment directed at tackling climate change, but are likely to deliver large benefits under relatively low risks.

Poverty trap
A poverty trap is a situation in which poverty is a self-reinforcing cycle. A poverty trap can be defined as an equilibrium outcome and a situation from which one cannot emerge without outside help, for example, via a positive windfall to a particular group, such as by redistribution or aid, or via a fundamental change in the functioning of markets.

Two mechanisms explaining poverty traps are common in the literature: the first one arises when resources fall short of a basic level: e.g. nutrition and health levels below which workers cannot effectively function or children cannot properly develop. A second mechanism, which has featured prominently, is the presence of market failures in the credit markets, which impedes efficient investments by the poor in both human and physical capital. (Source: Fuentes and Seck, 2008)

Regime
In the climate change context, the international climate regime generally refers to the institutions and decisions of the UN Framework Convention on Climate Change (**UNFCCC**). Regimes are social institutions consisting of agreed-upon principles, norms, rules, procedures, and programmes that govern the interactions of actors in specific issue areas (Levy *et al.*, 1995).

Resilience
The ability of a system and its component parts to anticipate, absorb, accommodate, or recover from the effects of a hazardous event in a timely and efficient manner, including through ensuring the

preservation, restoration, or improvement of its essential basic structures and functions (IPCC, 2012).

Stabilization

Keeping constant the atmospheric concentrations of one or more **greenhouse gases** (e.g. carbon dioxide) or of a CO_2-equivalent basket of greenhouse gases. Stabilization analyses or scenarios address the stabilization of the concentration of greenhouse gases in the atmosphere. (Source: IPCC, 2007)

Tipping points

A tipping point is defined in Alley *et al.* (2002) to 'occur when the climate system is forced to cross some threshold, triggering a translation to a new state at a rate determined by the climate system itself and faster than the cause'. The transition between states can happen suddenly and unexpectedly (as in the concept of 'abrupt climate change'), and be self-reinforcing and irreversible. The important point to note is that beyond a climate tipping the new equilibrium state for the climate system is beyond humans' control to determine. The presence of tipping points introduces a great deal of uncertainty in climate projections.

The presence of tipping points is characteristic of complex systems, such as the internet, the human brain or ecosystems. Such systems are dynamic and self-regulating. They have in-built mechanisms for self-correction that result in remarkable stability in the face of external stimuli. But they are also dynamic, meaning that their behavior is not constant or linear but subject to change. Complex systems can contain feedback loops that accelerate change.

As natural, climatic and socio-economic systems are increasingly interconnected, changes in one may lead to a tipping point in another. For example a natural tipping point, such as large-scale rainforest dieback, may be caused by a combination of climatic and human stressors, and in turn can accelerate climate change by releasing vast quantities of terrestrial carbon. Likewise, feedback loops in our socio-economic systems cause emissions to accelerate. For example, as more people use commercial airlines, the price of flying drops, allowing even more people to fly.

(Source: Alley, R.B. *et al.* (2002) *Abrupt Climate Change: Inevitable Surprises*, US National Research Council Report, National Academy Press, Washington, DC: Quotation from p. 14)

UNFCCC

The United Nations Framework Convention on Climate Change (UNFCCC; 'the Convention'), and its affiliated Kyoto Protocol is the most important international cooperation agreement on **climate change** and provides the basis for international action to curb global warming. The Convention was agreed in May 1992 and entered into force on March 21, 1994. The UNFCCC has been ratified by 194 countries, known as 'Parties to the Convention'. The Convention sets an overall framework for nations to cooperate to address climate change. (Source: Dawson and Spannagle, 2009)

Vulnerability

Vulnerability is the degree to which a system is susceptible to, and unable to cope with, adverse effects of **climate change**, including climate variability and extremes. Vulnerability is a function of the character, magnitude, and rate of climate change and variation to which a system is exposed, its sensitivity, and its **adaptive capacity**. (Source: IPCC, 2007)

Bibliography

Abbott, K.W. (2013) Constructing a Transnational Climate Change Regime: Bypassing and Managing States, (February 9, 2013).

Abdullah, A., Jallow, B.P. and Reazuddin, M. (2006) *Operationalising the Kyoto Protocol's Adaptation Fund: A New Proposal*, London: International Institute for Environment and Development.

Acemoglu, D. and Robinson, J.A. (2012) 'Why Nations Fail: The Origins of Power, Prosperity, and Poverty', New York: Crown Business Publications.

ACPC (2011) *Climate Science, Information, and Services in Africa: Status, Gaps and Policy Implications*, Working Paper 1, United Nations Economic Commission for Africa, Addis Ababa: African Climate Policy Centre.

ActionAid (2007) *Compensating for Climate Change: Principles and Lessons for Equitable Adaptation Funding*, Washington DC: ActionAid.

Adams, W.M. (2001) *Green Development: Environment and Sustainability in the Third World* (2nd edition), London: Routledge.

ADB (Asian Development Bank) (2005) 'Climate Proofing: A Risk-based Approach to Adaptation', Asian Development Bank Pacific Study Series, Manila, Philippines.

Adger, W.N., Arnell, N.W. and Tompkins, E.L. (2004) 'Successful adaptation to climate change across scales', *Global Environmental Change*, 15.2: 77–86.

Adger, W.N., Agrawala, S., Mirza, M.M.Q., Conde, C., O'Brien, K., Pulhin, J., Pulwarty, R., Smit, B. and Takahashi, K. (2007) 'Assessment of adaptation practices, options, constraints and capacity', in Parry, M.L., Canziani, O.F., Palutikof, J.P., van der Linden, P.J. and Hanson, C.E. (eds) (2007) *Climate Change 2007: Impacts, Adaptation and Vulnerability. Contribution of Working Group II to the Fourth Assessment Report of the Intergovernmental Panel on Climate Change*, Cambridge, UK, and New York, NY, USA: Cambridge University Press.

Adger, W.N., Paavola, Y., Huq, S. and Mace, M.J. (eds) (2006) *Fairness in Adaptation to Climate Change*, Cambridge, MA: MIT Press.

Adger, W.N., Dessai, S., Goulden, M., Hulme, M., Lorenzoni, I., Nelson, D., Naess, L.O., Wolf, J. and Wreford, A. (2009) 'Are there social limits to adaptation?', *Climatic Change*, 93: 335–354.

AfDB *et al.* (African Development Bank; Asian Development Bank; Department for International Development: United Kingdom; Directorate-General for International Cooperation: the Netherlands; Directorate General for Development: European Commission; Federal Ministry for Economic Cooperation and Development: Germany; Organization for Economic Cooperation and Development; United Nations Development Programme; United Nations Environment Programme; World Bank) (2003) *Poverty and Climate Change: Reducing the Vulnerability of the Poor Through Adaptation*, Washington DC: World Bank.

AGF (2010) *Report of the Secretary-General's High-level Advisory Group on Climate Change Financing*, 5 November 2010, New York: United Nations.

Agnew, C. and Woodhouse, P. (2010) *Water Resources and Development*, London: Routledge.

Agrawal, A. and Angelsen, A. (2012) 'Using community forest management to achieve REDD+ goals' in Angelsen, A., Brockhaus, M., Sunderlin, W.D. and Verchot, L.V. (eds) *Analysing REDD+: Challenges and choices*, Bogor: CIFOR.

Agrawala, S. (2005) *Bridge over Troubled Waters: Linking Climate Change and Development*, Paris: OECD.

Agrawala, S. and Fankhauser, S. (eds) (2008) *Economic Aspects of Adaptation to Climate Change. Costs, Benefits and Policy Instruments*, Paris: OECD.

Agrawala, S. and Van Aalst, M. (2008) 'Adapting development cooperation to adapt to climate change', *Climate Policy*, 8.2: 183–193.

Alam, K., Shamsuddoha, M., Tanner, T.M., Sultana, M., Huq, M.J. and Kabir, S.S. (2011) 'The Political Economy of Climate Resilient Development Planning in Bangladesh', *IDS Bulletin* 43.3: 52–61.

Allen, M.R., Mitchell, J.F.B. and Stott, P.A. (2013) 'Test of a decadal climate forecast', *Nature Geoscience* 6: 243–244.

Alley, R.B. *et al.*, (2002) *Abrupt Climate Change: Inevitable Surprises*, US National Research Council Report, Washington DC: National Academy Press.

Anderegg, W.R.L., Prall, J.W., Harold J. and Schneider, S.H. (2009) 'Expert credibility in climate change', *PNAS*, 107.27: 12107–12109.

Anderson, K. and Bows, A. (2008) 'Reframing the climate change challenge in light of post-2000 emission trends', *Phil. Trans. R. Soc. A*, 366: 3863–3882.

Andonova, L.B., Betsill, M. and Bulkeley, H. (2009) 'Transnational climate governance', *Global Environmental Politics*, 9.2: 52–73.

Andreae, M.O., Jones, C.J. and Cox, P.M. (2005) 'Strong present-day aerosol cooling implies a hot future', *Nature*, 435: 1187–1190.

Angelsen, A., Brockhaus, M., Sunderlin, W.D. and Verchot, L.V. (eds) (2012) *Analysing REDD+: Challenges and choices*, Bogor: CIFOR.

Angelsen, A. and McNeill, D. (2012) 'The evolution of REDD+' in Angelsen, A., Brockhaus, M., Sunderlin, W.D. and Verchot, L.V. (eds) *Analysing REDD+: Challenges and choices*, Bogor: CIFOR.

Atteridge, A., Kehler Siebert, C., Klein, R.J.T., Butler, C. and Tella, P. (2009) *Bilateral Finance Institutions and Climate Change: A Mapping of Climate Portfolios*, SEI Working Paper, Stockholm: Stockholm Environment Institute.

Ayers, J.M. and Huq, S. (2009) 'Supporting Adaptation to Climate Change: What Role for Official Development Assistance?', *Development Policy Review*, 27.6: 675–692.

Ayers, J., Schipper, E.L.F., Reid, H., Huq, S. and Rahman, A. (2013) *Community Based Adaptation to Climate Change: Scaling it up*, Abingdon: Routledge.

Bäckstrand, K. (2008) 'Accountability of Networked Climate Governance: The Rise of Transnational Climate Partnerships', *Global Environmental Politics*, 8.3: 74–102.

Baer, P., Athanasiou, T., Kartha, S. and Kemp-Benedict, E. (2009) *The Greenhouse Development Rights Framework: The right to development in a climate constrained world*, 2nd edition, Berlin: Heinrich-Böll-Stiftung.

Baeumler, A., Ijjasz-Vasquez, E. and Mehndiratta, S. (eds) (2012) *Sustainable Low-Carbon City Development in China*, Washington DC: World Bank.

Bahadur, A.V., Ibrahim, M. and Tanner, T.M. (2013) 'Characterising Resilience: Unpacking the concept for tackling climate change and development', *Climate and Development*, 5.1: 55–65.

Bailey, R. (2008) *Another inconvenient truth: How biofuel policies are deepening poverty and accelerating climate change*, Oxfam Briefing Paper 114, Oxfam International.

Ballesteros *et al.* (2010) 'Power, Responsibility, and Accountability: Re-Thinking the Legitimacy of Institutions for Climate Finance', Final Report. Washington DC: World Resources Institute.

Ban, K.M., Al-Madani, A., Barroso, J.M., Gurria, A., Kaberuka, D., Strauss-Kahn, D., Ping, J. and Zoellick, R. (2008), *Achieving the MDGs. Recommendations of the MDG Africa Steering Group* (New York), published by the United Nations Department of Public Information.

Bapna, M. and McGray, H. (2008) *Financing Adaptation: Opportunities for innovation and Experimentation*, WRI Conference Paper, Washington DC: World Resources Institute.

Barnett, J. and Adger, W.N. (2007) 'Climate Change, Human Security and Violent Conflict', *Political Geography*, 26.6: 639–655.

Barnett, J. and O'Neill, S. (2010) 'Maladaptation', Editorial, *Global Environmental Change* 20.2: 211–213.

Bausch, C. and Mehling, M. (2011) *Addressing the Challenge of Global Climate Mitigation: An Assessment of Existing Venues and Institutions*, Dialogue on Globalization Study, Berlin: Friedrich Ebert Stiftung.

Béné, C., Godfrey Wood, R., Newsham, A.J. and Davies, M. (2012) *Resilience: New Utopia or New Tyranny? Reflections about the Potentials and Limits of the Concept of Resilience in Relation to Vulnerability Reduction Programmes*, IDS Working Paper No. 405, Brighton: Institute of Development Studies.

Berg, A. and Ostry, J.D. (2011a) 'Equality and Efficiency: Is there a Tradeoff between the Two or Do the Two Go Hand in Hand?', *Finance and Development* 48.3: 12–15.

Berg, A. and Ostry, J.D. (2011b) *Inequality and Unsustainable Growth: Two Sides of the Same Coin?* IMF Staff Discussion Note 11/08, Washington DC: International Monetary Fund.

Bird, N. (2011) *The future for climate finance in Nepal*, case study for Capacity Development for Development Effectiveness Facility for Asia and Pacific, London: Overseas Development Institute.

Bird, N., Brown, J. and Schalatek, L. (2011a) *Design challenges for the Green Climate Fund*, Climate Finance Policy Brief No.4, North America: Heinrich Böll Foundation and London: ODI.

Bird, N., Billett, S. and Colon, C. (2011b) *Direct Access to Climate Finance: Experiences and lessons learned*, ODI/UNDP Discussion Paper, New York: UNDP.

Birkmann., J (ed.) (2006) *Measuring Vulnerability and Coping Capacity to Hazards of Natural Origin*, New York: United Nations University Press.

Bodansky, D. (2011) *Multilateral climate efforts beyond the UNFCCC*, C2ES Briefing, Arlington, VA: The Center for Climate and Energy Solutions.

Boyd, E., Hultman, N. E., Roberts, T., Corbera, E., Ebeling, J., Liverman, D.M., Brown, K., Tippmann, R., Cole, J., Mann, P., Kaiser, M., Robbins, M., Bumpus, A., Shaw, A., Ferreira, E., Bozmoski, A., Villiers, C. and Avis, J. (2007) *The Clean Development Mechanism: An Assessment of Current Practice and Future Approaches for Policy*, Working Paper 114, Norwich, UK: Tyndall Centre for Climate Change Research.

Bradley, R. and Baumert, K.A. (2005) *Growing in the Greenhouse – Protecting the Climate by Putting Development First*, Washington DC: World Resources Institute.

Braun, J., Swaminathan, M.S. and Rosegrant, M.W. (2005) *Agriculture, food security, nutrition and the Millennium Development Goals*, Washington: International Food Policy Research Institute (IFPRI).

Brooks, N. (2003) *Vulnerability, Risk and Adaptation: A Conceptual Framework*, Working Paper 38, Norwich, UK: Tyndall Centre for Climate Change Research.

Brooks, N., Adger, W.N. and Kelly, P.M. (2005) 'The determinants of vulnerability and adaptive capacity at the national level and the implications for adaptation', *Global Environmental Change*, 15: 151–163.

Brooks, N., Anderson, S., Ayers, J., Burton, I. and Tellam, I. (2011) *Tracking adaptation and measuring development*, London: International Institute for Environment and Development (IIED).

Brooks, N., Grist, N. and Brown, K. (2009) 'Development Futures in the Context of Climate Change: Challenging the Present and Learning from the Past', *Development Policy Review*, 27.6: 741–765.

Brown, J. (2009) *The future of climate finance: a new approach is needed*, ODI Opinion Background Note, London: Overseas Development Institute.

Brown, J., Bird, N. and Schalatek, L. (2010) *Climate finance additionality: emerging definitions and their implications*, Climate Finance Policy Brief No.2, North America: Heinrich Böll Foundation and London: ODI.

Brown, J. and Jacobs, M. (2011) *Leveraging private investment: the role of public sector climate finance*, ODI Background Note, London: Overseas Development Institute.

Brown, K. (2012) 'Policy discourses of resilience', in Pelling, M., Manuel-Navarrete, D. and Redclift, M. (eds) *Climate Change and the Crisis of Capitalism*, Abingdon: Routledge.

Bruinsma J. (ed.) (2003) *World Agriculture: Towards 2015/30, an FAO Perspective*, London: Earthscan and Rome: FAO.

Buchner, B., Brown, L. and Corfee-Morlot, J. (2011) *Monitoring and Tracking Long-term Finance to Support Climate Action*, COM/ENV/EPOC/IEA/SLT, Paris: OECD.

Bulkeley, H. and Newell, P.J. (2010) *Governing Climate Change*, Abingdon: Routledge.

Bulkeley, H. and Jordan, A. (2012) 'Transnational environmental governance: new findings and emerging research agendas', *Environment and Planning C: Government and Policy*, 30.4: 556–570.

Burniaux, J., Chateau, J. and Duval, R. (2010) *Is there a Case for Carbon-Based Border Tax Adjustment?: An Applied General Equilibrium Analysis*, OECD Economics Department Working Papers No. 794, Paris: OECD.

Burton, I., Huq, S., Lim, B., Pilifosova, O. and Schipper, E.L. (2002) 'From Impacts Assessment to Adaptation Priorities: The Shaping of Adaptation Policy', *Climate Policy*, 2: 145–159.

Burton, I. and May, E. (2004) 'The Adaptation Deficit in Water Resource Management', *IDS Bulletin*, 35.3: 31–37.

Butchart, S.H.M. *et al.* (2010) 'Global Biodiversity: Indicators of Recent Declines', *Science*, 328.5982: 1164–1168.

Byrne, R. Smith, A. Watson, J. and Ockwell, D. (2012) 'Energy Pathways in Low Carbon Development: The Need to Go Beyond Technology Transfer', in Ockwell D.G. and Mallet, A. (eds) *Low-Carbon Technology Transfer: From Rhetoric to Reality*, Abingdon: Routledge.

Cameron, C. (2011) Climate Change Financing and Aid Effectiveness: Ghana Case Study, Paris: OECD.

Cameron, E. (2013) *A Time for Leadership on Climate Justice*, WRI Insights, Washington DC: World Resources Institute.

Cannon, T., and Müller-Mahn, D. (2010) 'Vulnerability, resilience and development discourses in context of climate change', *Natural Hazards*, 55.3: 621–635.

Carter, T.R., Parry, M.L., Harasawa, H. and Nishioka, S. (1994) *IPCC Technical Guidelines for Assessing Climate Change Impacts and Adaptations*, Department of Geography, London, UK: University College London, and Japan: Centre for Global Environmental Research, National Institute for Environmental Studies.

CCCD (2010) *Closing the Gaps: Disaster risk reduction and adaptation to climate change in developing countries*, Report of the Commission on Climate Change and Development, Stockholm: Ministry for Foreign Affairs.

Challinor, A. (2008) 'Towards a Science of Adaptation that Prioritises the Poor', *IDS Bulletin* 39.4: 81–86.

Chambers, R. (1983) *Rural Development: Putting the Last First*, Essex: Longman Scientific & Technical.

Chandy, L. and Gertz, G. (2011) 'Poverty in numbers: the changing state of global poverty from 2005 to 2015', The Brookings Institution Policy Brief 2011–01, Washington DC: The Brookings Institution.

Chang, H-J. (2002) *Kicking Away the Ladder: Development Strategy in Historical Perspective*, London: Anthem Press.

Clapp J. and Dauvergne P. (2005) *Paths to a green world: the political economy of the global environment*, Cambridge MA: MIT Press.

Cline, W.R. (2007) *Global Warming and Agriculture: Impact Estimates by Country*, Washington DC: Center for Global Development and the Peterson Institute for International Economics.

Collins, K. and Ison, R. (2009) 'Jumping off Arnstein's ladder: Social learning as a new policy paradigm for climate change adaptation', *Environmental Policy and Governance*, 19.6: 358–373.

Comprehensive Assessment of Water Management in Agriculture (2007) 'Water for Food, Water for Life: A Comprehensive Assessment of Water Management in Agriculture', London: Earthscan, and Colombo: International Water Management Institute.

Conway, G.R. (1997) *The Doubly Green Revolution: Food for All in the 21st Century*, Cornell University Press.

Crutzen, P.J. (2002) 'Geology of mankind: the anthropocene', *Nature* 415.6867: 23.

Daly, H. (1996) *Beyond growth: The economics of sustainable development*, Boston: Beacon Press.

Daly, H. (1999) *Uneconomic growth in theory and in fact*, The First Annual Feasta Lecture, Trinity College, Dublin, 26th April.

Daly, H. (2005) 'Economics in a full world', *Scientific American*, 293.3: 100–107.

Davidson, O., Halsnaes, K., Huq, S., Kok, M., Metz, B., Sokona, Y. and Verhagen, J. (2003) 'The development and climate nexus: the case of sub-Saharan Africa', *Climate Policy*, 3: 97–113.

Davies, M., Guenther, B., Leavy, J., Mitchell T. and Tanner, T.M. (2009) *Climate Change Adaptation, Disaster Risk Reduction and Social Protection: Complementary Roles in Agriculture and Rural Growth?*, IDS Working Paper 320, Brighton: Institute of Development Studies.

Dawson, B. and Spannagle, M. (2009) *The Complete Guide to Climate Change*, Abingdon: Routledge.

Dazé, A., Ambrose, K. and Ehrhart, C. (2009) *Climate Vulnerability and Capacity Analysis Handbook*, Copenhagen: CARE International Poverty Environment and Climate Change Network.

De Lopez, T., Ponlok, T., Iyadomi, K., Santos, S., and McIntosh, B. (2009) 'Clean Development Mechanism and Least Developed Countries: Changing the Rules for Greater Participation', *The Journal of Environment and Development*, 18.4: 436–452.

Demetriades, J. and Esplen, E. (2008) 'The Gender Dimensions of Poverty and Climate Change Adaptation' *IDS Bulletin*, 35.3: 24–31.

Deneulin, S. and Shahani, L. (eds) (2009) *An Introduction to the Human Development and Capability Approach: Freedom and Agency*, London: Earthscan.

Depledge, J. and Yamin, F. (2010) 'The global climate change regime: A defence', in Helm, N. and Hepburn, C. (eds) *The Economics and Politics of Climate Change*, Oxford: Oxford University Press.

Dercon, S. and Christiansen, L. (2007) 'Consumption risk, technology adoption and poverty traps: evidence from Ethiopa', CSAE Working Paper Series, Oxford: Centre for the Study of African Economies, University of Oxford.

Dessai, S. and Hulme, M. (2004) 'Does climate adaptation policy need probabilities?' *Climate Policy*, 4: 107–128.

Dessai, S. and Wilby, R. (2012) 'How can developing country decision makers incorporate uncertainty about climate risks into existing planning and policymaking processes?' Washington DC: World Resources Report.

Dessler, A.E. (2012) *Introduction to Modern Climate Change*, Cambridge: Cambridge University Press.

Devereux, S. and Sabates-Wheeler, R. (2004) *Transformative Social Protection*, IDS Working Paper 232, Brighton: Institute of Development Studies.

DFID (2006) *Eliminating World Poverty: making governance work for the poor*, White Paper on International Development, London: UK Department for International Development.

DFID (2011) *Defining Disaster Resilience: A DFID Approach Paper*, London: UK Department for International Development.

Dietz, S., Anderson, D., Stern, N., Taylor, C. and Zenghelis, D. (2007) 'Right for the Right Reasons: A final rejoinder on the Stern Review', *World Economics*, 8.2: 229–258.

Disch, D. (2010) 'A comparative analysis of the "development dividend" of Clean Development Mechanism projects in six host countries', *Climate and Development*, 2: 50–64.

Dobbs, R., Oppenheim, J., Thompson, F., Brinkman, M. and Zornes, M. (2011) *Resource Revolution: Meeting the World's Energy, Materials, Food, and Water Needs*, New York: McKinsey Global Institute.

Dodman, D. (2009) 'Blaming cities for climate change? An analysis of urban greenhouse gas emissions inventories', *Environment and Urbanization*, 21.1: 185–201.

Drupp, M.A. (2011) 'Does the Gold Standard label hold its promise in delivering higher Sustainable Development benefits? A multi-criteria comparison of CDM projects', *Energy Policy*, 39.3: 1213–1227.

Dubash, N.K. (2009) *Climate Change through a Development Lens*, Background Paper, World Development Report 2010, Washington DC: World Bank.

Dunlap, R.E. and McCright, A.M. (2011) 'Organised climate change denial' in Dryzek, J.S., Norgaard, R.B. and Scholsberg, D. (eds) *The Oxford Handbook of Climate Change and Society*, Oxford: Oxford University Press.

Eakin, H.C. and Patt, A. (2011) 'Are adaptation studies effective, and what can enhance their practical impact?', *WIREs Climate Change*, 2: 141–153.

EC (2004) 'EU Strategy on Climate Change in the Context of Development Co-operation: Action Plan 2004–2008', Directorate General Environment, Brussels: European Commission.

Eriksen, S.H. and O'Brien, K. (2007) 'Vulnerability, poverty and the need for sustainable adaptation measures', *Climate Policy*, 7: 337–352.

Eriksen, S., Aldunce, P., Bahinipati, C.S., Martins, R.D., Molefe, J.I., Nhemachena, C., O'Brien, K., Olorunfemi, F., Park, J., Sygna, L. and Ulsrud, K. (2011) 'When not every response to climate change is a good one: identifying principles for sustainable adaptation', *Climate and Development*, 3: 7–20.

Esty, D.C. (2001) 'Toward Data-Driven Environmentalism: The Environmental Sustainability Index', *Environmental Law Reporter*, 31.5: 10603–10613.

Evans, A. (2010) *Globalisation and Scarcity: Multilateralism for a World with Limits*, New York: NYU Center on International Cooperation.

Evans, A. (2011) *Resource scarcity, fair shares and development*, WWF-UK/ Oxfam Discussion Paper, Oxford: Oxfam.

Evans, A., and Evans, J. (2011) *Resource Scarcity, Wellbeing and Development*, Commissioned Paper for the Bellagio Initiative, Brighton: Institute of Development Studies.

Evans, A. and Steven, D. (2012) *Sustainable development goals: A useful outcome from Rio+20*, New York: NYU CIC.

Fankhauser, S. (2010) 'The costs of adaptation', Wiley Interdisciplinary Reviews: Climate Change, 1: 23–30.

Fankhauser, S. and Burton, I. (2011) 'Spending adaptation money wisely', *Climate Policy*, 11: 1037–1049.

Fankhauser, S. and Schmidt-Traub, G. (2011) 'From adaptation to climate-resilient development: The costs of climate-proofing the Millennium Development Goals in Africa', *Climate and Development*, 3.2: 94–113.

FAO (2010) *Climate Smart Agriculture: Policies, Practices and Financing for Food Security, Adaptation and Mitigation*, Rome: FAO.

FAO (2011) *The State of Food Insecurity in the World 2011*, Rome: FAO.

FAO (2012) *Food Security and Climate Change*, Report of the High Level Panel of Experts on Food Security and Nutrition (HLPE), Rome: UN Committee on World Food Security (CFS).

Fay, M., Toman, M., Benitez, D. and Csordas, S. (2011) 'Infrastructure and sustainable development' in *Postcrisis Growth and Development: A Development Agenda for the G-20*, Washington DC: World Bank.

Foley, J. (2010) 'Boundaries for a Healthy Planet', *Scientific American*, 302.4: 54–57.

Folke, C. *et al.* (2011) 'Reconnecting to the Biosphere', *Ambio*, 40.7: 719–738.

Friedman, M. (1962) *Capitalism and Freedom*. Chicago: University of Chicago Press. (2002 edition).

Friend, R. and MacClune, K. (2012) 'Climate Resilience Framework: Putting Resilience into Practice', Boulder, USA: Institute for Social and Environmental Transformation – International.

Friends of the Earth (2011) 'Reckless Gamblers: How Politicians' Inaction is Ramping up the risk of dangerous climate change', Friends of the Earth, UK.

Fuentes, R. (2012) 'What kind of inequality matters most?: The case for unfairness', on Oxfam blog *From Poverty to Power*.

Fuentes-Nieva, R. and Pereira, I. (2010) 'The disconnect between indicators of sustainability and human development', in *Human Development Research Paper* 2010/34, United Nations Development Program.

Fuentes, R. and Seck, P. (2008) 'The Short and Long-Term Human Development Effects of Climate-Related Shocks: Some Empirical Evidence', Occasional Paper, *Human Development* Report Office.

Füssel, H.-M. (2010) 'How Inequitable is the Global Distribution of Responsibility, Capability, and Vulnerability to Climate Change: A Comprehensive Indicator-based Assessment', *Global Environmental Change*, 20.4: 597–611.

G8 (2005) Gleneagles Plan of Action, *Climate Change, Clean Energy and Sustainable Development*, London: UK G8 Presidency.

Gaffney, O., Seitzinger, S., Smith, M.S., Biermann, F., Leemans, R., Ingram, J., Bogardi, J., Larigauderie, A., Glaser, G., Diaz, S., Kovats, S., Broadgate, W., Morais, J. and Steffen, W. (2012) 'Interconnected risks and solutions for a planet under pressure: Transition to sustainability in the context of a green economy and institutional frameworks for sustainable development', Rio+20 Policy Brief, Planet Under Pressure Conference, March 26–29, London.

Gaillard, J.C. (2010) 'Vulnerability, capacity and resilience: Perspectives for climate and development policy', *Journal of International Development*, 22: 218–232.

Gallopín, G.C. (2006) 'Linkages between vulnerability, resilience and adaptive capacity', *Global Environmental Change*, 16.(2006): 293–303.

Gerst, M.D. and Raskin, P.D. (2011) *Turning Toward Sustainability: A Business-as-usual and an Alternative scenario prepared for the UN High-Level Panel on Global Sustainability*, Boston: Tellus Institute.

Girling, R. (2005) 'We're having a party', *Sunday Times* Magazine, July 3rd 2005.

Global Humanitarian Forum (2009) *Human Impact Report: Climate Change – The Anatomy of a Silent Crisis*, Geneva: Global Humanitarian Forum.

Gore, T. (2012) From superstorm Sandy to climate solidarity: How extreme weather can unlock climate action' on Oxfam blog *From Poverty to Power*.

Green, D. and Raygorodetsky, G. (2010) 'Indigenous knowledge of a changing climate', *Climatic Change*, 100: 239–242.

Green, D. (2012) 'Rio+20's sustainable development goals should reflect today's world', Poverty Matters blog, *The Guardian* online Friday 15th June 2012.

Grist, N. (2008) 'Positioning climate change in sustainable development discourse', *Journal of International Development*, 20.6: 783–803.

Grubb, M. (2004) 'Technology innovation and climate change policy: an overview of issues and options', *Keio Economic Studies*, 41.2: 103–132.

Gunderson, L. and Holling, C.S. (2001) *Panarchy: Understanding Transformations in Human and Natural Systems*, Washington DC: Island Press.

Gupta, J. (1997) *The Climate Change Convention and Developing Countries: From Conflict to Cooperation*, Dordrecht: Kluwer Academic Publishers.

Hagemann, M., Hendel-Blackford, S., Höhne, N., Harvey, B., Naess, L.O. and Urban, F. (2011), *Guiding climate compatible development: User-orientated analysis of planning tools and methodologies. Analytical report*, Cologne, Utrecht, London: Ecofys, and Brighton: Institute of Development Studies.

Haigh, C. and Vallely, B. (2010) *Gender and the Climate Change Agenda: The impacts of climate change on women and public policy*, London: Women's Environment Network.

Hammill, A. and Tanner, T.M. (2011) *Harmonising climate risk management: Adaptation screening and assessment tools for development co-operation*, OECD Environment Working Paper 36, ENV/WKP(2011)6, Paris: OECD.

Hammill, A., Matthew, R. and McCarter, E. (2008) 'Microfinance and Climate Change Adaptation', *IDS Bulletin*, 35.3: 113–122.

Hansen, J., Johnson, D., Lacis, A., Lebedeff, S., Lee, P., Rind, D. and Russell, G. (1981) 'Climate impact of increasing atmospheric carbon dioxide', *Science*, 213: 957–966.

Harmeling, S. and Kaloga, A. (2011) 'Understanding the political economy of the Adaptation Fund', *IDS Bulletin*, 42.3: 23–32.

Hart, S. and Prahalad, C.K. (2002) 'The Fortune at the Bottom of the Pyramid', *Strategy + Business*, 26: 54–67.

Hedger, M., Greeley, M. and Leavy, J. (2008) 'Evaluating Climate Change: Pro-Poor Perspectives', *IDS Bulletin*, 39.4: 75–80.

Heil, M., and Selden, T.M. (2001) 'International trade intensity and carbon emissions: a cross-country econometric analysis', *Journal of Environment and Development*, 10: 35–49.

Hellmuth, M.E., Moorhead, A., Thomson, M.C. and Williams, J. (eds) (2007) *Climate Risk Management in Africa: Learning from Practice*, International Research Institute for Climate and Society (IRI), New York: Columbia University.

Heltberg, R., Siegel, P.S. and Jorgensen, S.L. (2009) 'Addressing Human Vulnerability to Climate Change: Towards a "No-regrets" Approach', *Global Environmental Change*, 19: 89–99.

Hess, U. and Syroka, J. (2005) *Weather-based Insurance in Southern Africa: The Case of Malawi*, Agriculture and Rural Development Discussion Paper 13, Washington DC: World Bank.

Hibbard, K.A., Crutzen, P.J., Lambin, E.F., Liverman, D., Mantua, N.J., McNeill, J.R., Messerli, B. and Steffen, W. (2006) 'Decadal interactions of humans and the environment' in Costanza, R., Graumlich, L. and Steffen, W. (eds) *Sustainability or Collapse?: An Integrated History and Future of People on Earth*, Dahlem Workshop Report 96, Boston, MA: MIT Press, pp. 341–375.

Hinkel, J. (2011) 'Indicators of Vulnerability and Adaptive Capacity: Towards a Clarification of the Science–policy Interface', *Global Environmental Change*, 21.1: 198–208.

Hiraldo, R. and Tanner, T.M. (2011) *The global political economy of REDD+: Engaging social dimensions in the emerging green economy*, UNRISD Occasional Paper 4: Social Dimensions of Green Economy and Sustainable Development, Geneva: United Nations Research Institute for Social Development.

Hochrainer, S., Mechler, R., Pflug, G. (2007) 'Climate change and weather insurance in Malawi: assessing the impact', in Suarez, P., Linnerooth-Bayer, J. and Mechler, R. (eds) *Feasibility of Risk Financing Schemes for Climate Adaptation*, DEC Research Group, Infrastructure and Environment Unit, Washington DC: World Bank.

Hof, A.F., den Elzen, M.G.J. and Mendoza Beltran, A. (2011) 'Predictability, equitability and adequacy of post-2012 international climate financing proposals', *Environmental Science & Policy*, 14.6: 615–627.

Höhne, N., Vieweg, M., Hare, B., Schaeffer, M., Rogelj, J., Rocha, M., Gütschow, J., Schleussner, C-F., Fallasch, F., Larkin, J. and Fekete, H. (2012) *Warnings of climate science – again – written in Doha sand*, Climate Action Tracker Update, 8 December 2012, Bonn: Ecofys/Climate Analytics/PIK.

Ho-Lem, C., Zerriffi, H. and Kandlikar, M. (2011) 'Who participates in the Intergovernmental Panel on Climate Change and why: A quantitative assessment of the national representation of authors in the Intergovernmental Panel on Climate Change', *Global Environmental Change*, 21.4: 1308–1317.

Horn-Phathanothai, L. and Warburton, J. (2007) 'China's Environmental Crisis: What Does it Mean for Development?', *Development* 20.3: 2007.

Howes, M. (2005) *Politics and the Environment: Risk and the role of government and industry*, Sydney: Allen & Unwin/London: Earthscan.

Huq, S. and Reid, H. (2005) 'Mainstreaming adaptation into development', *IDS Bulletin*, 35.3: 15–21.

IFAD (2011) *Rural Poverty Report 2011*, Rome: IFAD.

International Monetary Fund (2007) *World Economic Outlook: Globalization and Inequality*, World Economic and Financial Surveys, Washington DC.

IPCC (1995a) *Climate Change 1995: IPCC Second Assessment – Summary for Policy Makers*, Geneva: IPCC.

IPCC (1995b) *A glossary by the Intergovernmental Panel on Climate Change*, Geneva: IPCC.

IPCC (2001a) *Climate Change 2001: Synthesis Report. A Contribution of Working Groups I, II, and III to the Third Assessment Report of the Intergovernmental Panel on Climate Change* [Watson, R.T. and the Core Writing Team (eds)], Cambridge, UK, and New York, NY, USA: Cambridge University Press.

IPCC (2001b) *Climate Change 2001: Mitigation Appendices. A Contribution of Working Group III to the Third Assessment Report of the Intergovernmental Panel on Climate Change.* Cambridge, UK, and New York, NY, USA: Cambridge University Press.

IPCC (2007) *Climate Change 2007: Synthesis Report. An Assessment of the Intergovernmental Panel on Climate Change*, Cambridge, UK, and New York, NY, USA: Cambridge University Press.

IPCC (2012) 'Summary for Policymakers', in Field, C.B., Barros, V., Stocker, T.F., Qin, D., Dokken, D.J., Ebi, K.L., Mastrandrea, M.D., Mach, K.J., Plattner, G.-K., Allen, S.K., Tignor, M. and Midgley, P.M. (eds), *Managing the Risks of Extreme Events and Disasters to Advance Climate Change Adaptation*, A Special Report of Working Groups I and II of the Intergovernmental Panel on Climate Change, Cambridge, UK, and New York, NY, USA: Cambridge University Press.

IPCC (2014) *Climate Change 2013: The Physical Science Basis. Summary for Policymakers*. Working Group I Contribution to the IPCC Fifth Assessment Report, Cambridge, UK, and New York, NY, USA: Cambridge University Press.

IRI (2006) *A Gap Analysis for the Implementation of the Global Climate Observing System Programme in Africa*, IRI Technical Report No. IRI-TR/06/1, Columbia: The International Research Institute for Climate and Society (IRI).

ISET (2008) *Re-imagining the Rural-Urban Continuum: Understanding the role ecosystem services play in the livelihoods of the poor in desakota regions undergoing rapid change*, Kathmandu: Institute for Social and Environmental Transition – Nepal (ISET-N).

Jackson, T. (2009) *Prosperity without growth: The transition to a sustainable economy*,: Sustainable Development Commission, Routledge.

Jackson, T. (2011) *Prosperity without growth: Economics for a finite planet*, Oxford and New York: Routledge.

Jacobs, M. (2013) 'Green Growth', in Falkner, R. (ed.), *Handbook of Global Climate and Environmental Policy*, Oxford: Wiley Blackwell.

Jacques, P.J., Dunlap, R.E. and Freeman, M. (2008) 'The Organization of Denial: Conservative Think Tanks and Environmental Skepticism', *Environmental Politics*, 17: 349–385.

Janssen, M.A. and Ostrom, E. (2006) 'Empirically based, agent-based models', *Ecology and Society*, 11: 37.

Jenson, J. (2010) *Defining and Measuring Social Cohesion*, London: Commonwealth Secretariat.

Johns, T. and Eyzaguirrea, P.B. (2007) 'Linking biodiversity, diet and health in policy and practice', symposium on 'Wild-gathered plants: basic nutrition, health and survival', *Proceedings of the Nutrition Society*, 65.2 (May 2006): 182–189.

Johnson, S. (2009) 'The Quiet Coup', *The Atlantic*, May 2009.

Jones, H. (2009) *Equity in development: why it is important and how to achieve it*, ODI Working Paper 311, Overseas Development Institute.

Jung, M., Eisbrenner, K. and Höhne, N. (2010) *How to get Nationally Appropriate Mitigation Actions (NAMAs) to work*, Ecofys Policy Update 11.2010, Cologne: Ecofys.

Kahn, M.E. (2005) 'The Death Toll from Natural Disasters: The Role of Income, Geography and Institutions', *The Review of Economics and Statistics*, 87.2: 271–284.

Kahneman, D. (2011) *Thinking Fast and Slow*, London: Allen Lane.

Kamal-Chaoui, L. and Robert, A. (eds) (2009) *Competitive Cities and Climate Change*, OECD Regional Development Working Papers No. 2, 2009, Paris: OECD.

Kareiva, P. (2010) 'Conservation Science: The End of the Wild?', interview with nature.org.

Kareiva, P., Watts, S., Mcdonald, R., and Boucher, T. (2007) 'Domesticated Nature: Shaping Landscapes and Ecosystems for Human Welfare', *Science*, 316.5833: 1866–1869.

Kelly, P.M. and Adger, W.N. (2000) 'Theory and practice in assessing vulnerability to climate change and facilitating adaptation', *Climatic Change*, 47.4: 325–352.

Keynes, J.M. (1964) *The General Theory*, New York: Harcourt Brace and World.

Khosla, V. (2011) *Black Swans thesis of energy transformation*, Khosla Ventures White Paper, August 2011.

Klare, M.T. (2012) *The Race for What's Left: The Global Scramble for the World's Last Resources*, New York: Metropolitan Books.

Klein, R.J.T. (1998) 'Towards better understanding, assessment and funding of climate adaptation', *Change*, 44: 15–19.

Klein, R.J.T. (2003) 'Adaptation to Climate Variability and Change: What is Optimal and Appropriate?' in Giupponi, C. and Schechter, M. (eds) *Climate Change and the Mediterranean: Socio-Economic Perspectives of Impacts, Vulnerability and Adaptation*, Cheltenham: Edward Elgar.

Klein, R.J.T., Huq, S., Denton, F., Downing, T.E., Richels, R.G., Robinson, J.B. and Toth, F.L. (2007) 'Inter-relationships between adaptation and mitigation' in Parry, L., Canziani, O.F., Palutikof, J.P., van der Linden, P.J. and Hanson, C.E. (eds) *Climate Change 2007: Impacts, Adaptation and Vulnerability. Contribution of Working Group II to the Fourth Assessment Report of the Intergovernmental Panel on Climate Change*, Cambridge, UK, and New York, NY, USA: Cambridge University Press.

Klein, R.J.T. (2009) 'Identifying Countries that are Particularly Vulnerable to the Adverse Effects of Climate Change: An Academic or a Political Challenge?', *Carbon & Climate Law Review* 3.3: 284–291.

Klein, R.J.T. and Möhner, A. (2011) 'The Political Dimension of Vulnerability: Implications for the Green Climate Fund', *IDS Bulletin* 42.3: 15–22.

Klein, R.J.T., Schipper, E.L.F., Dessai, S., (2005) 'Integrating mitigation and adaptation into climate and development policy: three research questions', *Environmental Science & Policy*, 8.6: 579–588.

Kok, M. *et al.* (2008) 'Integrating development and climate policies: National and international benefit', *Climate Policy*, 8.2: 103–118.

Kolbert, E. (2011) 'Enter the Anthropocene: Age of Man' in *National Geographic*, 219: 60–77.

Korten, D.C. (1990) *Getting to the 21st Century: Voluntary Action and the Global Agenda*, West Hartford, CT: Kumarian Press.

Kuznets, S. (1955) 'Economic Growth and Income Inequality', *American Economic Review*, 45.1: 1–28.

Kuznets, S. (1962) 'How to Judge Quality', *The New Republic*, 20 October: 29–32.

Laan, T., Beaton, C. and Presta, B. (2010) 'Strategies for Reforming Fossil-Fuel Subsidies: Practical lessons from Ghana, France and Senegal', The Global Subsidies Initiative, *Untold Billions: Fossil-fuel Subsidies, Their Impact and the Path to Reform*, Geneva: IISD.

Lal, R. (2011) 'Sequestering carbon in soils of agro-ecosystems', *Food Policy*: doi 10.1016/j.foodpol.2010.12.001.

LDC Expert Group (2009) *Support needed to fully implement national adaptation programmes of action (NAPAs)*, Bonn: UNFCCC.

LDC Expert Group (2013) *National Adaptation Plans: Technical guidelines for the national adaptation plan process*, Bonn: UNFCCC.

Leach, M. (ed.) (2008) *Re-framing resilience: A symposium report*, STEPS Working Paper 13, Brighton: Institute of Development Studies.

Lee, B., Preston, F., Kooroshy, J., Bailey, R. and Lahn, G. (2012) *Resources Futures*, London: Chatham House.

Legros, G., Havet, I., Bruce, N. and Bonjour, S. (2009) *The Energy Access Situation in Developing Countries: A Review Focusing on the Least Developed Countries and Sub-Saharan Africa*, New York: UNDP and World Health Organization.

Leichenko, R. and O'Brien, K. (2008) *Double Exposure*, New York: Oxford University Press.

Lenton, T.M., Held, H., Kriegler, E., Hall, J.W., Lucht, W., Rahmstorf, S. and Schellnhuber, H.J. (2008) 'Tipping Elements in the Earth's Climate System', *Proceedings of the National Academy of Sciences*, 105.6: 1786–1793.

Levina, E. (2007) *Adaptation to Climate Change: International Agreements for Local Needs*. Document prepared by the OECD and IEA for the Annex I Expert Group on the UNFCCC. Paris: OECD/IEA.

Levy, M.A., Young, O.R. and Zürn, M. (1995) 'The Study of International Regimes', *European Journal of International Relations*, 1.3: 267–330.

Linnerooth-Bayer, J. and Mechler, R. (2006) 'Insurance for assisting adaptation to climate change in developing countries: A proposed strategy', *Climate Policy*, 6: 621–663.

Lockwood, M. and Cameron, C. (2011) *Low Carbon Energy and Development: Bridging Concepts and Practice for Low Carbon Climate Resilient Development*, Learning Hub Bridging Paper 3, Brighton: Institute of Development Studies.

Lomborg, B. (2001) *The Skeptical Environmentalist: Measuring the Real State of the World*, Cambridge University Press.

Lovelock, J. (2010) *The Vanishing Face of Gaia: A Final Warning*, London: Penguin.

Lu, F. *et al.* (2009) 'Soil carbon sequestrations by nitrogen fertilizer application, straw return and no-tillage in China's cropland', *Global Change Biology*, 15: 281–305.

Ludwig, F., Kabat, P., van Shaik, H. and van der Valk, M. (eds) (2009) *Climate Change Adaptation in the Water Sector*, London: Earthscan.

Luxbacher, K. and Goodland, A. (2010) 'Building Resilience to Extreme Weather: Index-Based Livestock Insurance in Mongolia', Case Studies written for the World Resources Report 2011, Washington DC: WRI.

Lynas, M. (2011) *The God Species*, London: Fourth Estate.

McGray, H., Bradley, R. and Hammill, A. (2007) *Weathering the Storm: Options for Framing Adaptation and Development*, Washington DC: World Resources Institute.

McKinsey & Co. (2009a) *Pathways to a Low-Carbon Economy. Version 2 of the Global Greenhouse Gas Abatement Cost Curve*, New York: McKinsey & Company.

McKinsey & Co. (2009b) *Shaping Climate Resilient Development: a Framework for Decision Making*, Report of the Economics of Adaptation Working Group, Geneva and New York: Swiss Re and McKinsey & Company.

°Magee, T. (2012) *A Field Guide to Community Based Adaptation*, London: Earthscan.

Mahul, O. and Skees, J. (2007) *Managing Agricultural Risk at the Country Level: The Case of Index-Based Livestock Insurance in Mongolia*, Policy Research Working Paper 4325, Washington DC: World Bank.

Mastrandrea, M.D., Field, C.B., Stocker, T.F., Edenhofer, O., Ebi, K.L., Frame, D.J., Held, H., Kriegler, E., Mach, K.J., Matschoss, P.R., Plattner, G.-K., Yohe, G.W. and Zwiers, F.W. (2010) *Guidance Notes for Lead Authors of the IPCC Fifth Assessment Report on Consistent Treatment of Uncertainties*, Intergovernmental Panel on Climate Change, Geneva: IPCC.

Matsuo, N. (2003) 'CDM in the Kyoto negotiations: How CDM has worked as a bridge between developed and developing worlds?', *Mitigation and Adaptation Strategies for Global Change*, 8.3: 191–200.

Meadows, D. L., Randers, J. and Behrens III, W.W. (1972) *The Limits to Growth: A Report for the Club of Rome's Project on the Predicament of Mankind.* New York: Universe Books.

Mechler, R., Linnerooth-Bayer, J. and Peppiatt, D. (2006) *Disaster Insurance for the Poor?: A Review of Microinsurance for Natural Disaster Risks in Developing Countries*, Geneva and Laxenburg: ProVention and the International Institute for Applied Systems Analysis.

Meze-Hausken, E., Patt, A. and Fritz, S. (2009) 'Reducing climate risk for micro-insurance providers in Africa: A case study of Ethiopia', *Global Environmental Change*, 19.1: 66–73.

Mignone, B.K., Socolow, R.H., Sarmiento, J.L. and Oppenheimer, M. (2008) 'Atmospheric stabilization and the timing of carbon mitigation', *Climatic Change*, 88.3–4: 251–265.

Milanovic, B. (2010) *The Haves and the Have-Nots: A Brief and Idiosyncratic History of Global Inequality*, New York: Basic Books.

Mill, J.S. (1848) *The Principles of Political Economy with some of their Applications to Social Philosophy*, London: Longmans, Green and Co.

Millennium Ecosystem Assessment (2005) *Ecosystems and Human Well-being*, Washington DC: World Resources Institute.

Mitchell, D. (2008) *A note on Rising Food Prices*, World Bank Policy Research Working Paper No. 4682, Washington DC: World Bank.

Mitchell, T. and Van Aalst, M. (2008) *Convergence of Disaster Risk Reduction and Climate Change Adaptation: A Review for DFID*, Brighton: Institute of Development Studies.

Mitchell, T. and Van Aalst, M. (2011) *Headlines from the IPCC Special Report on Extreme Events*, Overseas Development Institute and Red Cross/ Red Crescent Climate Centre.

Mitchell, T.C. and Maxwell, S. (2010) *Defining climate compatible develop-ment*, CDKN Policy Brief, London: Climate and Development Knowledge Network.

Mitchell, T.C. and Harris, K. (2012) *Resilience: A risk management approach*, ODI Background Note, London: Overseas Development Institute.

Mitlin, D. and Hickey, S. (2000) *Rights-based Approaches to Development: Exploring the Potential and Pitfalls*, Sterling, VA: Kumarian Press,

Mohieldin, M. and Allaoua, Z. (2013) *Getting Cities Right*, Commentary, New York: Project Syndicate.

Möhner, A. and Klein, R.J.T. (2007) *The Global Environment Facility: Funding for adaptation or adapting to funds?*, Climate and Energy Programme Working Paper, Stockholm: Stockholm Environment Institute.

Moncel, R., Joffe, P., McCall, K. and Levin, K. (2011) *Building the Climate Change Regime: A Survey and Analysis of Approaches*, Working Paper, Washington DC: World Resources Institute.

Morgan, J. (2011) 'Filling the Sustainability Innovation Gap', *The EcoInnovator* blog, 10th October 2011.

Moser, S.C. (2008) *Resilience in the Face of Global Environmental Change*, CARRI Research Paper No. 2, prepared for Oak Ridge National Laboratory and its Community and Regional Resilience Initiative (CARRI), Oak Ridge, TN: ORNL.

Moser, S.C. and Boykoff, M.T. (eds) (2013) *Successful Adaptation to Climate Change: Linking Science and Practice in a Rapidly Changing World*, Abingdon: Routledge.

Müller, B. and Hepburn, C. (2006) *IATAL – An Outline Proposal for an International Air Travel Adaptation Levy, EV36*, Oxford Institute for Energy Studies, EV 36, October, Oxford: Oxford Institute for Energy Studies.

Müller, B. (2008) *International Adaptation Finance: The Need for an Innovative Strategic Approach*, Oxford: Oxford Institute for Energy Studies.

Müller, B. (2009) *Additionality in the Clean Development Mechanism: Why and What?*, Working Paper EV 44, Oxford: Oxford Institute for Energy Studies.

Müller, B. (2011) *Time to Roll Up the Sleeves – Even Higher!: Longer-term climate finance after Cancun*, Oxford: Oxford Energy and Environment Brief.

Müller, B. and Khan, F.I. (2011) *The Green Climate Fund: What needs to be done for Durban (COP 17)* Oxford Energy and Environment Brief.

Nabuurs, G.J. *et al.* (2007) 'Forestry', in Metz, B., Davidson, O.R., Bosch, P.R., Dave, R. and Meyer, L.A. (eds) *Climate Change 2007: Mitigation. Contribution of Working Group III to the Fourth Assessment Report of the Intergovernmental Panel on Climate Change*, Cambridge, UK and New York, USA: Cambridge University Press.

Najam, A. (2005) 'Developing Countries and Global Environmental Governance: From Contestation to Participation to Engagement', *International Environmental Agreements*, 5.3: 303–321.

Najam, A., Rahman, A.A., Huq, S. and Sokona, Y. (2003) 'Integrating sustainable development into the Fourth Assessment Report of the Intergovernmental Panel on Climate Change', *Climate Policy*, 3S1: S9–S17.

Najam, A. *et al.* (2007) 'Sustainable Development and Mitigation', in Metz, B., Davidson, O.R., Bosch, P.R., Dave, R. and Meyer, L.A. (eds) *Climate Change 2007: Mitigation. Contribution of Working Group III to the Fourth Assessment Report of the Intergovernmental Panel on Climate Change*, Cambridge, UK and New York, USA: Cambridge University Press.

Narain, U., Margulis, S. and Essam, T. (2011) 'Estimating costs of adaptation to climate change', *Climate Policy*, 11.3: 1001–1019.

National Intelligence Council (2012) *Global Trends 2030: Alternative Worlds*, Washington DC: US Government Printing Office.

Nellemann, C. and Corcoran, E. (eds) (2010) *Dead Planet, Living Planet – Biodiversity and Ecosystem Restoration for Sustainable Development. A Rapid Response Assessment*, Arendal, Norway: United Nations Environment Programme, GRID-Arendal.

Nelson, D.R., Adger, W.N. and Brown, K. (2007) 'Adaptation to Environmental Change: Contributions of a Resilience Framework', *Annual Review of Environment and Resources*, 32: 395–419.

Nerlich, C. (2010) ' "Climategate": Paradoxical metaphors and political paralysis', *Environmental Values*, 19.4: 419–442.

New, M., Liverman, D., Schroeder, H. and Anderson, K. (2011) 'Four degrees and beyond: The potential for a global temperature increase of four degrees and its implications', *Philosophical Transactions of the Royal Society of London A*, 369: 6–19.

New Scientist (2011) 'The biggest climate change uncertainty of all', *New Scientist*, 212.2835: 1–5.

Nilsson, A.E. and Swartling, A.G. (2009) *Social Learning about Climate Adaptation: Global and Local Perspectives*, SEI Working Paper, Stockholm: Stockholm Environment Institute.

Niosi, J. (2010) *Building National And Regional Innovation Systems*, London: Edward Elgar.

Norse, D. (2012) 'Low carbon agriculture: Objectives and policy pathways', in *Environmental Development*, 1.1: 25–39.

NRDC (2012) 'Governments Should Phase Out Fossil Fuel Subsidies or Risk Lower Economic Growth, Delayed Investment in Clean Energy and Unnecessary Climate Change Pollution', in *Fuel Facts*, Natural Resources Defense Council.

O'Brien K., Eriksen, S., Nygaard, L., Schjolden, A. (2007) 'Why different interpretations of vulnerability matter in climate change discourses', *Climate Policy*, 7.1: 73–88.

O'Brien, K. (2009) 'Do values subjectively define the limits to climate change adaptation?' in O'Brien, K., Adger, W.N. and Lorenzoni, I. (eds) *Adapting to climate change: Thresholds, values, governance*, Cambridge: Cambridge University Press.

O'Brien, K., Hayward, B. and Berkes, F. (2009) 'Rethinking social contracts: building resilience in a changing climate', *Ecology and Society*, 14.2: Art. 12.

O'Brien, K. (2012) 'Global environmental change II: From adaptation to deliberate transformation', *Progress in Human Geography*, 36.5: 667–676.

Ockwell, D.G., Watson, J., MacKerron, G., Pal, P. and Yamin, F. (2008) 'Key policy considerations for facilitating low carbon technology transfer to developing countries', *Energy Policy*, 36.11: 4104–4115.

Ockwell, D.G. and Mallet, A. (2012) 'Introduction: Low-Carbon Technology Transfer – from Rhetoric to Reality' in Ockwell, D.G. and Mallet, A. (eds) *Low-Carbon Technology Transfer: From Rhetoric to Reality*, Abingdon: Routledge.

OECD (2006) *Declaration on Integrating Climate Change Adaptation into Development Co-operation*, Adopted by Development and Environment

Ministers of OECD Member Countries, 4 April 2006, Paris: Organisation for Economic Co-operation and Development.

OECD (2009) *Policy Guidance on Integrating Climate Change Adaptation into Development Co-operation*, Paris: Organisation for Economic Co-operation and Development.

OECD (2010a) *Cities and Climate Change*, Paris: Organisation for Economic Co-operation and Development.

OECD (2010b) *Development Perspectives for a Post-2012 Climate Financing Architecture*, Paris: Organisation for Economic Co-operation and Development.

OECD (2012a) 'Development aid: Grants by private voluntary agencies', *Development: Key Tables from OECD*, No. 3: doi: 10.1787/aid-pvt-vol-table-2012-1-en.

OECD (2012b) *Inclusive Green Growth: For the Future We Want*, Paris: OECD.

OECD/ IEA (2011) *World Energy Outlook 2011*, Paris: IEA Publication Service.

Okun, A. (1975) *Equality and Efficiency: The Big Tradeoff*, Washington: Brookings Institution Press.

Olbrisch, S., Haites, E., Savage, M., Dadhich, P. and Shrivastava, M.K. (2011) 'Estimates of incremental investment for and cost of mitigation measures in developing countries', *Climate Policy*, 11.3: 970–986.

Oxfam GB, India (2004) *Riding the Storm: Community Experiences from Disaster Preparedness Initiatives in Andhra Pradesh*, Delhi: Oxfam GB, India.

Oxfam International (2007) *Adapting to Climate Change: What's Needed in Poor Countries and Who Should Pay*, Oxfam Briefing Paper 104, Washington DC, Brussels, Geneva and New York: Oxfam.

Oxfam (2012) 'Our land, our lives: Time out on the global land rush', Oxfam Briefing Note, October 2012.

Pacala, S. and Socolow, R. (2004) 'Stabilization Wedges: Solving the Climate Problem for the Next 50 Years with Current Technologies', *Science*, 305.5686: 968–972.

Page, E. (2006) *Climate Change and Future Generations*, Cheltenham: Edward Elgar.

Pan, J. (2005) 'Meeting Human Development Goals with Low Emissions: An Alternative to Emissions Caps for post-Kyoto from a Developing Country Perspective', *International Environmental Agreements* 5: 89–104.

Park, S.E., Marshall, N.A., Jakku, E., Dowd, A.M., Howden, S.M., Mendham, E., Fleming, A. (2012) 'Informing adaptation responses to climate change through theories of transformation', *Global Environmental Change*, 22: 115–126.

Parker, C. and Mitchell, A. (2009) *The Little REDD+ Book: A guide to governmental and non-governmental proposals for reducing emissions from deforestation and forest degradation*, Oxford: Global Canopy Programme.

Parry, M., Rosenzweig, C., Iglesias, A., Livermore, M. and Fischer, G. (2004) 'Effects of climate change on global food production under SRES emissions and socio-economic scenarios', *Global Environmental Change*, 14.1: 53–67.

Parry, M., Canziani, O.F., Palutikof, J.P. *et al.* (2007) 'Technical Summary', in Parry, M., Canziani, O.F., Palutikof, J.P., van der Linden, P.J. and Hanson, C.E. (eds) *Climate Change 2007: Impacts, Adaptation and Vulnerability. Contribution of Working Group II to the Fourth Assessment Report of the Intergovernmental Panel on Climate Change*, Cambridge, UK, and New York, NY, USA: Cambridge University Press.

Parry, M., Arnell, N., Berry, P., Dodman, D., Fankhauser, S., Hope, C., Kovats, S., Nicholls, R., Satterthwaite, D., Tiffin, R. and Wheeler, T. (2009) *Assessing the Costs of Adaptation to Climate Change: A Review of the UNFCCCC and Other Recent Estimates*, London: International Institute for Environment and Development and the Grantham Institute for Climate Change, Imperial College.

Patz, J.A., Gibbs, H.K., Foley, J.A., Rogers, J.V. and Smith, K.R. (2007) 'Climate change and global health: quantifying a growing ethical crisis', *EcoHealth* 4: 397–405.

Pelling, M. (2010) *Adaptation to climate change: From resilience to transformation*, Oxford, UK: Routledge.

Persson, A. (2011) *Institutionalising climate adaptation finance under the UNFCCC and beyond: Could an adaptation 'market' emerge?*, SEI Working Paper No. 2011–03, Stockholm: Stockholm Environment Institute.

Peters, G.P., Marland, G., Le Quéré, C., Boden, T., Canadell, J.G. and Raupach, M.R. (2012) 'The challenge to keep global warming below 2 °C', *Nature Climate Change*, 3: 4–6.

Pielke, R. Jr (2010) *The Climate Fix: What scientists and politicians won't tell you about global warming*, New York: Basic Books.

Pouliotte, J., Smit, B. and Westerhoff, L. (2009) 'Adaptation and development: Livelihoods and climate change in Subarnabad, Bangladesh', *Climate and Development*, 1: 31–46.

Pritchett, L. (1997) 'Divergence, Big Time', in *The Journal of Economic Perspectives*, 11.3: 3–17.

Rabinovitch, J. and Leitman, J. (1993) *Environmental innovation and management in Curitiba, Brasil*, Washington DC: UNDP/UNCHS/World Bank.

Raupach, M.R. *et al.* (2007) 'Global and regional drivers of accelerating CO_2 emissions', *Proceedings of the National Academy of Sciences of the United States of America*, (PNAS) 104.24: 10288–93.

Ravallion, M., Heil, M. and Jalan, J. (2000) 'Carbon emissions and income inequality', *Oxford Economic Papers*, 52.4: 651–669.

Raworth, K. (2012) *A Safe and Just Space for Humanity: Can we live within the doughnut?*, Oxfam Discussion Paper, February 2012, Oxford: Oxfam.

Roberts, J.T. and Parks, B.C. (2007) *A Climate of Injustice: Global Inequality, North-South Politics and Climate Policy*, Cambridge, MA: MIT Press.

Roberts, J.T. *et al.* (2010): *Copenhagen's climate finance promise: six key questions*, London: IIED.

Roberts, R.A.J. (2005) 'Insurance of Crops in Developing Countries', FAO Agricultural Services Bulletin 159, Rome: FAO.

Rockström, J., Steffen, W., Noone, K., Persson, Å., Chapin, F.S., Lambin, E.F., Lenton, T.M. and Foley, J.A. (2009) 'A safe operating space for humanity', *Nature*, 461.7263: 472–475.

Rose, S.K. *et al.* (2011) 'Land-based mitigation in climate stabilization', in *Energy Economics*, 34.1: 365–380.

SABMiller (2012) *SABMiller Position Paper – Energy and Carbon: Reducing our energy and carbon footprint*, www.sabmiller.com/files/pdf/positionpaper_energy.pdf

Satterthwaite, D. (2009) 'The implications of population growth and urbanization for climate change', in Paper presented at Expert-Group Meeting on Population

Dynamics and Climate Change, UNFPA and IIED in collaboration with UN-HABITAT and the Population Division, UN/DESA.

Save the Children (2011) *Severe Child Poverty: Nationally and Locally*, Save the Children Briefing, February 2011.

Schalatek, L. and Bird, N. (2011) *The Principles and Criteria of Public Climate Finance – A Normative Framework*, Climate Finance Fundamentals, Brief 1, Washington DC: Heinrich Böll Foundation and London: ODI.

Schalatek, L., Bird, N. and Brown, J. (2010) *Where's the Money? The Status of Climate Finance Post-Copenhagen. The Copenhagen Accord, UNFCCC Negotiations and a Look at the Way Forward*, North America: Heinrich Böll Foundation, and London: ODI.

Schalatek, L., Nakhooda, S. and Bird, N. (2012) *The Green Climate Fund*, Climate Finance Fundamentals, Brief 11, North America: Heinrich Böll Foundation, and London: ODI.

Schipper, E.L.F. (2006) 'Conceptual history of adaptation in the UNFCCC process', *Review of European Community & International Environmental Law*, 15.1: 82–92.

Schipper, L. and Pelling, M. (2006) 'Disaster risk, climate change and international development: scope for, and challenges to, integration', *Disasters*, 30.1: 19–38.

Schipper, E.L. and Burton, I. (2009b) 'Understanding Adaptation: Origins, Concepts, Practice, Policy', in Schipper, E.L. and Burton, I. (eds) *Earthscan Reader on Adaptation to Climate Change*, London: Earthscan.

Schmalensee, R., Stoker, T.M. and Judson, R.A. (1998) 'World Carbon Dioxide Emissions: 1950–2050', *The Review of Economics and Statistics*, 80.1: 15–27.

Schneider, S.H. (2009) *Science as a Contact Sport: Inside the Battle to Save Earth's Climate*, Washington DC: National Geographic Society.

Seballos, F. and Kreft, S. (2011) 'Towards an Understanding of the Political Economy of the PPCR', *IDS Bulletin*, 42.3: 33–41.

Sen, A. (1999) *Development as Freedom*, New York: Anchor.

Shahani, L. and Deneulin, S. (2011) *An Introduction to the Human Development and Capability Approach: Freedom and Agency*, London: Routledge.

Shue, H. (1993) 'Subsistence Emissions and Luxury Emissions', *Law & Policy*, 15.1: 39–60.

Shue, H. (1999) 'Global Environment and International Inequality', *International Affairs*, 75.3: 531–545.

Silva Villanueva, P. (2011) *Learning to ADAPT: Monitoring and evaluation approaches in climate change adaptation and disaster risk reduction – challenges, gaps and ways forward*, SCR Working Paper 9, Brighton: Institute of Development Studies.

Simon, J. (1983) *The Ultimate Resource*, Princeton, NJ: Princeton University Press.

Singer, P. (2007) *One World: The Ethics of Globalisation*, New Haven: Yale University Press.

Smit, B. and Wandel, J. (2006) 'Adaptation, adaptive capacity and vulnerability', *Global Environmental Change*, 16.3: 282–292.

Smit, B., Burton, I., Klein, R.J.T. and Wandel, J. (2000) 'An Anatomy of Adaptation to Climate Change and Variability', *Climatic Change*, 45.1: 223–251.

Smit, B., Pilifosova, O., Burton, I., Challenger, B., Huq, S., Klein, R.J.T. and Yohe, G. (2001) 'Adaptation to climate change in the context of sustainable development and equity' in McCarthy, J.J. Canziani, O., Leary, N.A., Dokken, D.J. and White, K.S. (eds) *Climate Change 2001: Impacts, Adaptation and Vulnerability. Contribution of the Working Group II to the Third Assessment Report of the Intergovernmental Panel on Climate Change*, Cambridge, United Kingdom and New York, NY, USA: Cambridge University Press.

Smith, P. *et al.* (2007) 'Agriculture', in Metz, B., Davidson, O.R., Bosch, P.R., Dave, R. and Meyer, L.A. (eds) *Climate Change 2007: Mitigation. Contribution of Working Group III to the Fourth Assessment Report of the Intergovernmental Panel on Climate Change.* Cambridge, United Kingdom and New York, NY, USA: Cambridge University Press.

Smith, P. *et al.* (2007) 'Policy and technological constraints to implementation of greenhouse gas mitigation options in agriculture', in *Agriculture, Ecosystems & Environment*, 118.1–4: 6–28.

Sorrell, S. (2007) *The rebound effect: An assessment of the evidence for economy-wide energy savings from improved energy efficiency*, London: UK Energy Research Centre.

Speth, G.J. (2008) *Bridge at the Edge of the World: Capitalism, the Environment, and Crossing from Crisis to Sustainability*, New Haven: Yale University Press.

Stadelmann, M., Roberts, J.T. and Huq, S. (2010) *Baseline for Trust: Defining 'New and Additional' Climate Funding*, IIED Briefing Paper, London: International Institute for Environment and Development.

Stafford Smith, M., Horrocks, L., Harvey, A. and Hamilton, C. (2011) 'Rethinking adaptation for a 4°C world', *Philosophical Transactions of the Royal Society of London A*, 369: 196–216.

Steffen, W., Persson, Å., Deutsch, L., Zalasiewicz, J., Williams, M., Richardson, K., Crumley, C. *et al.* (2011) 'The Anthropocene: From Global Change to Planetary Stewardship', *Ambio*, 40.7: 739–761.

Stern, N. (2007) *The Economics of Climate Change: The Stern Review*, Cambridge: Cambridge University Press.

Sterner, T. (2003) *Policy Instruments for Environmental and Natural Resource Management*, Washington DC: Resources for the Future.

Stiglitz, J., Sen, A. and Fitoussi, J.-P. (2009) Report by the Commission on the Measurement of Economic Performance and Social Progress, SSRN Electronic Journal: doi:10.2139/ssrn.1714428.

Stiglitz, J. (2010) 'Needed: a new economic paradigm', *Financial Times*, August 19, 2010.

Stiglitz, J. (2012) *The Price of Inequality: The Avoidable Causes and Invisible Costs of Inequality*, London: Allen Lane.

Stirling, A. (2003) 'Risk, Uncertainty and Precaution: Some Instrumental Implications from the Social Sciences', in Berkhout, F., Leach, M. and Scoones, I. (eds), *Negotiating Environmental Change: New Perspectives from Social Science*, London: Edward Elgar.

Suarez, P., Linnerooth-Bayer, J., Mechler, R. (eds) (2007) *Feasibility of Risk Financing Schemes for Climate Adaptation*, DEC Research Group, Infrastructure and Environment Unit, Washington DC: The World Bank.

Suarez, P. and Linnerooth-Bayer, J. (2010) 'Micro Insurance for Local Adaptation', *WIREs Climate Change*, 1.2: 271–278.

Suarez, P. and Linnerooth-Bayer, J. (2011) *Insurance-related instruments for disaster risk reduction*, Input paper for the 2011 Global Assessment Report on Disaster Risk Reduction, Geneva: International Strategy for Disaster Reduction (UNISDR).

Sukhdev, P., Wittmer, H., Schröter-Schlaack, C., Nesshöver, C., Bishop, J., ten Brink, P., Gundimeda, H., Kumar, P., Simmons, B. and UNEP Ginebra (Suiza) (2010) *The economics of ecosystems and biodiversity: Mainstreaming the economics of nature: a synthesis of the approach, conclusions and recommendations of TEEB*, The Economics of Ecosystems and Biodiversity.

Sumner, A. (2012) 'Where do the Poor Live?', *World Development*, 40.5: 865–877.

Swanstrom, T. (2008) *Regional Resilience: A Critical Examination of the Ecological Framework*, USA: Institute of Urban and Regional Development.

Tanner, T.M. and Mitchell, T. (2008) 'Introduction: Building the Case for Pro-Poor Adaptation', *IDS Bulletin*, 39.4: 1–5.

Tanner, T.M. and Mitchell, T. (2008) 'Entrenchment or Enhancement: Could Climate Change Adaptation Help Reduce Chronic Poverty?', *IDS Bulletin*, 39.4: 6–15.

Tanner, T.M., Garcia, M., Lazcano, J., Molina, F., Molina, G., Rodríguez, G., Tribunalo, B. and Seballos, F. (2009) 'Children's participation in community-based disaster risk reduction and adaptation to climate change', *Participatory Learning and Action*, 60: 54–64.

Tanner, T.M. and Allouche, J. (2011) 'Towards a new political economy of climate change', *IDS Bulletin* 43.3: 1–14.

Tanner, T.M., Lockwood, M. and Seballos, F. (2012) *Learning to Tackle Climate Change*, Brighton: Institute of Development Studies.

Tanner, T.M. and Bahadur, A.V. (2013) *Distilling the characteristics of transformational change in a changing climate*, Paper presented at the conference Transformation in a Changing Climate, University of Oslo, 19–21st June.

Tawney, L., Almendra, F., Torres, P. and Weischer, L. (2011) *Two Degrees of Innovation: How to Seize the Opportunities in Low-carbon Power*, WRI Working Paper, Washington DC: World Resources Institute.

The Government Office for Science (2011) *Foresight: Migration and Global Environmental Change – Future Challenges and Opportunities*, Final Project Report, London: The Government Office for Science.

Thornton, N. (2011) *Realising Development Effectiveness: Making the Most of Climate Change Finance in Asia and the Pacific. A synthesis report from five country studies in Bangladesh, Cambodia, Indonesia, Philippines and Vietnam*, CDDE Report, London: Agulhas Applied Knowledge.

Thrupp, L.A. (2000) 'Linking Agricultural Biodiversity and Food Security: the Valuable Role of Agrobiodiversity for Sustainable Agriculture', *International Affairs*, 76.2: 283–297.

Tietenberg, T.H. and Lewis, L. (2000) *Environmental and natural resource economics*, Reading, MA: Addison-Wesley.

Toman, M. (2006) 'Values in the economics of climate change', *Environmental Values*, 15.3: 365–379.

Tompkins, E.L., Nicholson-Cole, S.A., Hurlston, L.A., Boyd, E., Hodge, G.B., Clarke, J. Gray, G., Trotz, N. and Varlack, L. (2005) *Surviving Climate Change in Small Islands: A Guidebook*, Norwich, UK: Tyndall Centre for Climate Change Research.

Transparency International (2010) *Global Corruption Report: Climate Change*, London: Earthscan.

Tschakert, P. and Dietrich, K.A. (2010) 'Anticipatory learning for climate change adaptation and resilience', *Ecology and Society*, 15.2: 11.

UNDP (1990) *Human Development Report 1990: Concept and Measurement of Human Development*, New York: Oxford University Press.

UNDP (2005) *Adaptation policy frameworks for climate change. Developing strategies, policies and measures*, Cambridge and New York: Cambridge University Press.

UNDP (2007) *Human Development Report 2007/2008 Fighting Climate Change: Human Solidarity in a Divided World*, New York: United Nations Development Programme.

UNDP (2009) *Charting a low-carbon route to development*, New York: United Nations Development Programme.

UNDP and UNFCCC (2009) *Handbook for Conducting Technology Needs Assessment for Climate Change*, New York: United Nations Development Programme.

UNDP (2011) *Human Development Report 2011 – Sustainability and Equity: A Better Future for All*, New York: Palgrave Macmillan.

UNEP (2011a) *Bridging the Emissions Gap*, United Nations Environment Programme (UNEP), Nairobi: UNEP.

UNEP (2011b) *Towards a Green Economy – Pathways to Sustainable Development and Poverty Eradication – A Synthesis for Policy Makers*, UNEP.

UNFCCC (1992) *United Nations Framework Convention on Climate Change*, Bonn: UNFCCC.

UNFCCC (2006) *Technologies for adaptation to climate change*, Bonn: UNFCCC.

UNFCCC (2007a) *Investment and financial flows to address climate change*, Bonn: UNFCCC.

UNFCCC (2007b) *Report on existing and potential investment and financial flows relevant to the development of an effective and appropriate international response to climate change*, Bonn: UNFCCC.

UNFCCC (2008) *Investment and financial flows to address climate change: an update*, Technical Paper, FCCC/TP/2008/7, Bonn: UNFCCC.

UNFCCC (2011) *Green Climate Fund – report of the Transitional Committee: Draft Decision -/CP.17*, 10 December 2011.

UNFCCC (2012) *Report of the Conference of the Parties on its eighteenth session, held in Doha from 26 November to 8 December 2012. Addendum Part Two: Action taken by the Conference of the Parties at its eighteenth session*, FCCC/CP/2012/8/Add1, Bonn: UNFCCC.

UNGA (2012) *The future we want*, Resolution adopted by the United Nations General Assembly, Document A/RES/66/288, New York: United Nations.

UNISDR (2008) *Climate change and disaster risk reduction*, Briefing Note 1, Geneva: UN International Strategy for Disaster Reduction.

United Nations Population Division (2010) *World Urbanisation Prospects: the 2009 Revision Population Database*, UNDESA.

US Census Bureau (2012) 'World POPClock Projection', July 2012–July 2013 data.

UVEK (2008) *Funding Scheme for Bali Action Plan: A Swiss Proposal for Global Solidarity in Financing Adaptation*, 'Bali Paper' Updated for SB28 Bonn, Bern: Federal Office for the Environment.

van Aalst, M. (2006) *Managing Climate Risk: Integrating Adaptation into World Bank Group Operations*, Global Environment Facility Program, Washington DC: World Bank.

van Aalst, M., Burton, I. and Cannon, T. (2008) 'Community level adaptation to climate change: the potential role of participatory community risk assessment', *Global Environmental Change*, 18.1: 165–179.

van den Bergh, J.C.J.M. (2009) 'The GDP paradox', *Journal of Economic Psychology*, 30.2: 117–135.

van der Brugge, R. and Rotmans, J. (2007) 'Towards transition management of European water resources', *Water Resource Management*, 21: 249–267.

van Vuuren, D., Hoogwijk, M., Barker, T., Riahi, K., Boeters, S., Chateau, J., Scrieciu, S., van Vliet, J., Masui, T., Blok, K., Blomen, E. and Kram, T. (2009) 'Comparison of top-down and bottom-up estimates of sectoral and regional greenhouse gas emission reduction potentials', *Energy Policy*, 37: 5125–5139.

Venugopal, S. and Srivastava, A. (2012) *Moving the Fulcrum: A Primer on Public Climate Financing Instruments Used to Leverage Private Capital*, Washington DC: World Resources Institute.

Wade, R. (2003) 'What strategies are viable for developing countries today? The World Trade Organization and the shrinking of "development space" ', *Review of International Political Economy*, 10.4: 627–644.

Ward, P. and Shively, J. (2011) 'Vulnerability, Income Growth and Climate Change', *World Development*, 3.

Warner, K., Ranger, N., Surminski, S., Arnold, M., Linnnerooth-Bayer, J., Michel-Kerjan, E., Kovacs, P. and Herweijer, C. (2009) *Adaptation to Climate Change: Linking Disaster Risk Reduction and Insurance*, Geneva: United Nations International Strategy for Disaster Reduction.

Warner, K. and Zakieldeen, S.A. (2012) *Loss and damage due to climate change: An overview of the UNFCCC negotiations*, Oxford: European Capacity Building Initiative (ECBI).

Watson, C., Nakhooda, S., Caravani, A. and Schalatek, L. (2012) *The practical challenge of monitoring climate finance: Insights from Climate Funds Update*, Climate Finance Policy Brief, North America: Heinrich Böll Foundation and London: ODI.

Watson, R.T. *et al.* (2000) *Land Use, Land Use Change, and Forestry*, Special Report of the Intergovernmental Panel on Climate Change, Geneva: IPCC.

WBCSD (2010) *Vision 2050: The new agenda for business*, Geneva: World Business Council of Sustainable Development.

WCED (World Commission on Environment and Development) (1987) *Our Common Future* (Brundtland Report), Oxford: Oxford University Press.

Weart, S. (2008) *The Discovery of Global Warming*, 2nd edition, Cambridge, MA: Harvard University Press.

Whitmarsh, L. (2011) 'Scepticism and uncertainty about climate change: dimensions, determinants and change over time', *Global Environmental Change*, 21.2: 690–700.

WHO/WMO (2012) *Atlas of Climate and Health*, Geneva: Joint publication by the World Health Organization and the World Meteorological Organization.

Wilby, R.L., Troni, J., Biot, Y., Tedd, L., Hewitson, B.C., Smith, D.G. and Sutton, R.T. (2009) 'A review of climate risk information for adaptation and development planning', *International Journal of Climatology*, 29: 1193–1215.

Wilby, R.L. and Dessai, S. (2010) 'Robust adaptation to climate change', *Weather*, 65: 180–185.

Wilkinson, R.G., and Pickett, K. (2010) *The spirit level: Why greater equality makes societies stronger*, New York: Bloomsbury Press.

Willis, K. (2005) *Theories and Practices of Development*, New York and London: Routledge.

Wisner, B., Blaikie, P., Cannon, T. and Davis, I. (2004) *At risk: Natural hazards, people's vulnerability and disasters*, London: Routledge.

Wood, S., Ericksen, P., Stewart, B., Thornton, P. and Anderson, M. (2010) 'Lessons Learned from International Assessments' in Ingram, J., Erickson, P., and Liverman, D. (eds) *Food Security and Global Environmental Change*, London, UK and Washington DC: Earthscan.

World Bank (2006) *Investment Framework for Clean Energy and Development*, Washington DC: World Bank.

World Bank (2008) 'Rising Food and Fuel Prices: Addressing the Risks to Future Generations', Human Development Network (HDN) and Poverty Reduction and Economic Management (PREM) Network, Washington DC: World Bank.

World Bank (2010a) *World Development Report 2010: Development and Climate Change*, Washington DC: World Bank.

World Bank (2010b) *Economics of Adaptation to Climate Change: Synthesis Report*, Washington DC: World Bank.

World Bank (2010c) *Cities and Climate Change: An Urgent Agenda*, Washington DC: World Bank.

World Bank (2011) *Toward a Partnership for Sustainable Cities*, Washington DC: World Bank.

World Bank (2012a) *The State and Trends of the Carbon Market 2012*, Washington DC: World Bank.

World Bank and OECD (2012b) *Integrating Human Rights into Development: Donor Approaches, Experiences and Challenges*, The Development Dimension, Washington DC: World Bank.

World Bank. (2012c). *Inclusive Green Growth: The Pathway to Sustainable Development*, Washington DC: World Bank.

World Bank (2012d) *Turn down the heat: Why a 4°C warmer world must be avoided*, A Report for the World Bank by the Potsdam Institute for Climate Impact Research and Climate Analytics, Washington DC: World Bank.

World Economic Forum (2012) 'Global Risks 2012', An Initiative of the Risk Response Network, 7th edition, Cologny/Geneva: World Economic Forum.

World Food Programme (WFP) and the International Fund for Agricultural Development (IFAD) (2010), *The Potential for Scale and Scaleability in Weather Index Insurance*, Geneva: IFAD.

World Resources Institute (WRI) in collaboration with United Nations Development Programme, United Nations Environment Programme, and World Bank (2005) *World Resources 2005 – The Wealth of the Poor: Managing Ecosystems to Fight Poverty*, Washington DC: WRI.

World Resources Institute (WRI) in collaboration with United Nations Development Programme, United Nations Environment Programme, and World Bank (2011) *World Resources 2010–2011 – Decision Making in a Changing Climate: Adaptation Challenges and Choices*, Washington DC: WRI.

Wrathall, D., Oliver-Smith, A., Sakdapolrak, P., Gencer, E., Feteke, A. and Lepana-Reyes, M. (2013) *Problematising loss and damage*, UNU-EHS working paper, Bonn: United Nations University.

WWF (2012) *2012 Living Planet Report: Biodiversity, biocapacity and better choices*, Geneva: WWF International.

Yamin, F. and Depledge, J. (2004) *The International Climate Change Regime: A guide to Rules, Institutions and Procedures*, Cambridge: Cambridge University Press.

Young, Z. (2002) *A New Green Order? The World Bank and the Politics of the Global Environment Facility*, London: Pluto Press.

Zolli, A. and Healy, A.M. (2012) *Resilience: Why Things Bounce Back*, London: Headline.

Index